Titans of the Climate

American and Comparative Environmental Policy
Sheldon Kamieniecki and Michael E. Kraft, series editors

For a complete list of books in the series, please see the back of the book.

Titans of the Climate

Explaining Policy Process in the United States and China

Kelly Sims Gallagher and Xiaowei Xuan

The MIT Press
Cambridge, Massachusetts
London, England

This book was set in ITC Stone Serif Std by Toppan Best-set Premedia Limited. Printed and bound in the United States of America.

Library of Congress Cataloging-in-Publication Data

Names: Gallagher, Kelly Sims, author. | Xuan, Xiaowei, author.
Title: Titans of the climate : explaining policy process in the United States and China / Kelly Sims Gallagher and Xiaowei Xuan ; forewords by John P. Holdren and Junkuo Zhang.
Description: Cambridge, MA : The MIT Press, [2018] | Series: American and comparative environmental policy | Includes bibliographical references and index.
Identifiers: LCCN 2018010218| ISBN 9780262038751 (hardcover : alk. paper) | ISBN 9780262535847 (pbk. : alk. paper)
Subjects: LCSH: Climatic changes--Government policy--United States. | Climatic changes--Government policy--China. | Greenhouse gas mitigation--Government policy--United States. | Greenhouse gas mitigation--Government policy--China. | Carbon dioxide mitigation--Government policy--United States. | Carbon dioxide mitigation--Government policy--China.
Classification: LCC QC903.2.U6 G35 2018 | DDC 363.738/745610951--dc23 LC record available at https://lccn.loc.gov/2018010218

10 9 8 7 6 5 4 3 2 1

Contents

Series Foreword

Kelly Sims Gallagher and Xiaowei Xuan's book on climate change policy in the United States and China could not be more timely given the changes taking place within each nation and globally. In 2015, the United States accounted for 15 percent and China for 29 percent of global greenhouse gas emissions, more than other nations in the world. On a per capita basis, the United States was significantly higher than China in its emissions. When the G7 leaders were unsuccessful in persuading President Donald Trump to keep the United States in the Paris Climate Accord in May 2017, French President Emmanuel Macron turned to the leaders of the other nations and said, "China leads." Many might argue that America's exit from the accord marks the end of the nation's leadership on climate change. However, given the overwhelming evidence of the severity of the problem and its very serious negative effects on the world, it is hard to imagine that future presidents will embrace President Trump's position. In fact, despite President Trump's decision to pull out of the Paris Climate Accord, numerous American states and cities are adopting policies to reduce their greenhouse gas emissions, and many leading U.S. corporations remain supportive of the agreement.

Until the rejection of the accord by the United States, many if not most observers were pointing to China as the main obstacle to developing and implementing necessary stringent reductions in greenhouse gas emissions. Whether true or not, China now has compelling domestic reasons to change its energy consumption profile from a greater emphasis placed on the use of fossil fuels, particularly coal, to a greater use of nonfossil fuels, especially cleaner renewable sources of energy such as wind and solar power. Among other reasons, a desire to reduce air pollution in large cities such as Beijing and Shanghai has convinced many in China's national government of the immediate need to push for the adoption of cleaner fuels.

There is still internal resistance in the country by some powerful interests, such as the state-owned energy companies, to the formulation of policies designed to reduce the use of coal and other fossil fuels and to the pursuit of a low-carbon economic model. No doubt, a number of prominent Chinese government officials see the nation's future leadership position in the global climate debate as a way to convince others in government and industry of the importance of reducing greenhouse gas emissions and improving its stature around the world. In fact, China already leads the world in installed capacity of wind and solar power, a feat that would not have been possible without the global climate change debate and the need to develop a low-carbon economy. How all this plays out in the future will be fascinating to observe.

Gallagher and Xuan's book focuses on how the United States and China develop and implement climate change policy domestically. The book begins with a discussion of the national circumstances within each country concerning energy resources and consumption, sources of greenhouse gas emissions, economic growth and development, and the policy landscape of climate change. Gallagher and Xuan then compare the policymaking structures, actors, processes, and approaches in the United States and China. Chapters on the formulation of target greenhouse gas emissions reductions and their implementation within each nation follow. The authors then analyze why climate policy outcomes differ across the two countries. An excellent conclusion summarizes their observations and provides important insights into the present and likely future policy approaches of the two largest greenhouse gas emitters in the world.

Gallagher and Xuan decided to write a comparative book so that there would be two major points of reference for students of climate change policy. As they observe, climate policy in the United States and China is like a mosaic, and how analysts view policymaking in each nation will depend on which nation they know best. If readers are more familiar with the American system, then they will more easily be able to understand the points of departure in the Chinese system, and vice versa. The goal of the authors is, through a dispassionate analysis, to help readers better understand the policy approaches of the country they know least about, especially the challenges, constraints, and opportunities of that country. Although this book focuses on climate policy, students will also learn a great deal about the broader contemporary policy processes in both countries.

The book illustrates well our purposes in the MIT Press series in American and Comparative Environmental Policy. We encourage work that examines a broad range of environmental policy issues. We are particularly interested in volumes that incorporate interdisciplinary research and focus on the linkages between public policy and environmental problems and issues both within the United States and in cross-national settings. We welcome contributions that analyze the policy dimensions of relationships between humans and the environment from either a theoretical or empirical perspective.

At a time when environmental policies are increasingly seen as controversial and new and alternative approaches are being implemented widely, we especially encourage studies that assess policy successes and failures, evaluate new institutional arrangements and policy tools, and clarify new directions for environmental politics and policy. The books in this series are written for a wide audience that includes academics, policymakers, environmental scientists and professionals, business and labor leaders, environmental activists, and students concerned with environmental issues. We hope they contribute to public understanding of environmental problems, issues, and policies of concern today and also suggest promising actions for the future.

Sheldon Kamieniecki, University of California, Santa Cruz
Michael Kraft, University of Wisconsin-Green Bay
Coeditors, American and Comparative Environmental Policy Series

Foreword by John P. Holdren

The joint leadership of the United States and China played a pivotal role in bringing the nations of the world together to sign the December 2015 Paris climate agreement—a groundbreaking set of national commitments and side agreements establishing, for the first time, a truly global process for reducing the greenhouse gas emissions that are disrupting the world's climate. Notwithstanding President Trump's misguided determination to withdraw from the Paris Agreement, the fact is that strengthening and deepening the US-China leadership that led to the agreement will be essential if that global compact is to succeed as the necessary next step toward avoiding climate catastrophe. *Titans of the Climate* provides policymakers, scholars, and advocates for stabilizing the global climate a rich understanding of the climate policy processes of these two countries, whose continued leadership in this domain is so critical to the success of the Paris Agreement and the needed steps beyond.

The long-term goal stated in the Paris Agreement is to "hold the increase in global average temperature to well below 2°C above pre-industrial levels and pursue efforts to limit the temperature increase to 1.5°C." The Intended Nationally Determined Contributions (INDCs) of the 195 signatory nations, which do not extend beyond 2030, will not suffice to meet either target. Indeed, they are not even sufficient to get the world onto an optimal trajectory for meeting either one by 2030. That is why the agreement calls for countries to revisit their commitments every five years after 2015 with an eye to increased ambition, and why still deeper cuts after 2030 will be needed even if such ambition materializes and is realized in the form of actual emission cuts.[1]

Although other anthropogenic greenhouse gases contribute to the problem, CO_2 has been responsible for about 60 percent of the cumulative warming influence of human emissions since the start of the Industrial Revolution and is responsible for about 80 percent of the future warming being entrained by current human emissions. Of this 80 percent, 85–90 percent comes from the so-called industrial sources (primarily fossil fuel burning and secondarily cement manufacturing) and most of the remainder from agriculture and land-use change. Thus, the "industrial" emissions of CO_2 are the single largest driver of the global problem—and, remarkably, China and the United States together were responsible in 2016 for 43 percent of these.[2] That is why many informed observers have been saying for decades that unless and until the United States and China—the largest industrial nation emitter and the largest developing nation emitter, respectively—agreed to lead a global effort in emissions reductions, the rest of the world could hardly be expected to follow.

That agreement happened in November 2014, when President Obama and President Xi stood up together in Beijing and announced their Joint Statement on Climate Change, acknowledging the two countries' dominant roles in driving climate change and committing to specific targets and timetables for emissions reductions. It happened, after years of disagreements and misunderstanding, because of the growing recognition of the need to act on the part of both governments and a deepening understanding of the other country's decision processes, policies, and interests. Kelly Sims Gallagher and Xiaowei Xuan have been leaders in building that understanding in the two countries and between them, and they were personally involved behind the scenes in the run-up to the Beijing announcement. It is difficult to imagine two people better positioned to have written this book.

Titans of the Climate provides the fascinating history, contemporary understanding, and map for the future that can and should be the basis of continuing joint leadership by the United States and China in addressing the global climate change challenge. Policymakers, business leaders, scholars, and students from both countries should take note.

John P. Holdren is the Teresa and John Heinz Professor of Environmental Policy at the Kennedy School of Government; codirector of the program on Science, Technology, and Public Policy in the Kennedy School's Belfer

Center for Science and International Affairs; and professor of environmental science and policy in the Department of Earth and Planetary Sciences at Harvard University. From January 2009 to January 2017, he was President Obama's science advisor and the Senate-confirmed director of the White House Office of Science and Technology Policy (OSTP), becoming the longest-serving science advisor to the president in the history of the position.

Foreword by Junkuo Zhang

Climate change is a colossal challenge for humankind; it affects the well-being of the world's people and the future prospects of all nations. Recent more frequent climate extremes, including heat waves, droughts, and floods, have shown that climate change characterized by global warming has a significant impact on natural ecosystems throughout the world and seriously threatens the survival and development of the human condition. It is clearly now a very pressing task for each country to determine how to mitigate climate change and improve its ability to effectively adapt to climate change.

As the largest developing country, China is undergoing rapid industrialization and urbanization and is confronted with the multiple tasks of economic development, poverty reduction, and environmental protection. To be a responsible country, China has taken its own initiative to address climate change, making efforts to control greenhouse gas emissions and improving its ability to adapt to climate change. China takes seriously its responsibilities to deeply participate in global governance, build a community of human destinies, and promote common development for all mankind.

China not only recognizes its international obligations and responsibilities but also believes it is an inherent requirement to achieve sustainable development in its domestic economy and society. In so doing, China will actively address climate change and promote green and low-carbon development. In recent years, China has undertaken a series of policy measures that include promotion of energy conservation and energy-efficiency standards, development and deployment of renewable energy, expansion of forest carbon sinks, establishment of a nationwide carbon emissions trading market, creation of low-carbon and adaptation regional pilots, mobilization of green finance, and passage of climate change legislation. With these policies, China has made remarkable achievements in promoting energy conservation, emission reduction, and low-carbon development. China has achieved a marked reduction in the intensity of greenhouse gas emissions, and it has realized positive improvements in environmental quality, energy structure, and industrial structure. China's experience shows that promoting low-carbon transformation will help to address climate change, improve the ecological environment, and promote the sustained development of the economy. While promoting green development, a circular economy, and low-carbon development vigorously, and building a beautiful homeland, China has made positive contributions to the global efforts to address climate change.

China and the United States are the world's first- and second-largest emitters of greenhouse gases. Without good cooperation between the two countries, we cannot address global climate change effectively. When President Xi Jinping met with President Trump in April 2017, he commented, "We have a thousand reasons to improve Sino-US relations and there is no single reason to ruin Sino-US relations." Although the recent changes in the Paris Agreement have introduced some uncertainties to Sino-US climate cooperation, the general trend of green, low-carbon global development will not change and the tide of international cooperation on climate change will not be reversed.

This book, coauthored by Professor Kelly Sims Gallagher and Senior Researcher Xuan Xiaowei, provides a detailed analysis of the climate policies of China and the United States. More commendably, through a dispassionate analysis the two authors help readers "put themselves in the other's shoes" to better understand, ideally with some empathy, the challenges, constraints, and opportunities of each country in the climate arena. This book is of great value to readers to deepen their understanding about the climate policies of China and the United States and to promote climate cooperation between the two countries in related fields.

Junkuo Zhang
Vice President, the Development Research Center of the State Council (DRC), China

Junkuo Zhang is the vice president (vice minister) and a senior research fellow of the DRC. He has done extensive research on macroeconomics, regional economy, and economic reform, was awarded a government special allowance in 1997, and won the Sun Yefang Economics Prize in 1998 and the Special Prize of China Development Research in 2005 and 2013. The DRC is a top think tank in China and directly under the State Council. Its major function is to conduct research and provide policy options and consulting advice to the CPC Central Committee and the State Council.

序 言

气候变化是全人类面临的重大挑战,关系到世界人民福祉和各国长远发展。近期在世界范围内频繁出现的热浪、干旱和洪水等极端气候事件表明,以全球变暖为主要特征的气候变化,对全球自然生态系统产生显著的影响,给人类的生存和发展带来了严峻的挑战,如何有效应对气候变化、提高适应气候变化能力已成为各国十分现实和紧迫的任务。

中国作为最大的发展中国家,正处在工业化、城镇化快速发展的阶段,面临着发展经济、消除贫困、保护环境等多重挑战。作为负责任的大国,中国政府把积极应对气候变化、努力控制温室气体排放、提高适应气候变化的能力作为深度参与全球治理、打造人类命运共同体和推动全人类共同发展的责任担当。

积极应对气候变化,推动绿色低碳发展,既是中国作为发展中大国应尽的国际义务和责任,也是中国实现国内经济社会可持续发展的内在要求。近年来,中国采取了节能和提高能效、发展可再生能源、增加森林碳汇、建立全国碳排放权交易市场、低碳试点示范、发展绿色金融、推进气候变化立法、气候适应型城市建设试点等一系列政策措施,在推动节能减排和低碳发展等领域取得了明显的成绩,不仅实现了温室气体排放强度的显著下降,而且还在环境质量、能源结构和产业结构等方面产生了积极的变化。中国的经验证明,大力推进能源低碳转型,不仅有助于应对气候变化和改善生态环境,也会促进经济持续长期有效发展。中国在大力推进绿色发展、循环经济、低碳发展,努力建设美丽家园的同时,也为全球应对气候变化做出了积极的贡献。

中国和美国作为世界温室气体排放的第一和第二大国,有效应对全球气候变化问题,离不开两国的良好合作。正如2017年4月习近平主席与特朗普总统会晤时所指出:"我们有一千条理由把中美关系搞好,没有一条理由把中美关系搞坏"。尽管近期的新变化给中美气候合作进程带来一些不确定性,但全球绿色低碳循环发展的大趋势不会改变,国际合作应对气候变化的潮流也不会逆转。

加拉格尔(Gallagher)教授和宣晓伟研究员合著的这本书,对中美两国的气候变化政策做了详尽的分析,更难能可贵的是,两位研究者分别站在自身和对方的立场,换位思考、推己及人,深入地探讨了两国在制定和实施气候变化政策时面临的约束、挑战和机遇。本书对于读者加深对中美两国气候政策的理解,推进两国在相关领域的合作,有着重要的价值。是为序。

张军扩 研究员
国务院发展研究中心 副主任

张军扩,国务院发展研究中心副主任,研究员;长期从事宏观经济、区域经济、经济改革方面研究工作,荣获1997年政府特殊津贴、1998年第八届孙冶方经济科学奖和中国发展研究奖特等奖(2005和2013年)。国务院发展研究中心是直属中国国务院的政策研究和咨询机构,主要职能是研究中国国民经济、社会发展和改革开放中的全局性、战略性、前瞻性、长期性问题,为党中央和国务院提供政策建议和咨询意见。

Acknowledgments

We would like to begin by acknowledging one another's contributions because this has truly been a collaborative effort from start to finish. We both learned through our conversations, exchanges of drafts, challenging questions posed to each other, and the continuous effort to clarify similarities and differences.

Our families deserve recognition for supporting this endeavor because we had to spend time away from them in order to work together on this book. Both of us have sons nearly the same age, and Kelly also has a daughter, and our families have now lived in each other's countries and met each other. We hope that our effort to demystify the policy processes of our two countries will help their generation to foster mutual understanding in pursuit of peace and cooperation.

Special thanks to those who served as invaluable research assistants for this book: Qi Qi (齐琦) and Chris Sall, both graduate students at the Fletcher School. Both conducted literature reviews in both Chinese and English, and Qi Qi (齐琦) contributed significantly to the policy inventory. Jillian DeMair provided editorial assistance. For those who took the time to comment on parts or all of this book, we warmly thank you for making it better. In alphabetical order, we thank anonymous reviewers, Kevin Gallagher, John P. Holdren, Ken Lieberthal, Gregory Nemet, Qi Ye (齐晔), and Fang Zhang (张芳). Kelly also presented an early version of the book in a presentation at the University of Cambridge at the invitation of Laura Diaz Anadon. To Beth Clevenger, Anthony Zannino, and Clay Morgan at the MIT Press, we have appreciated your encouragement, patience, and counsel.

We also want to acknowledge our current and former colleagues in the government who helped us understand how things work on both sides of the Pacific. All remaining errors, of course, are our own.

List of Acronyms

BLM—US Bureau of Land Management

CBRC—China Banking Regulatory Commission (中国银行业监督管理委员会 or 银监会)

CCP—Chinese Communist Party (中国共产党)

CH₄—Methane

CMA—China Meteorological Administration (中国气象局)

CO₂—Carbon dioxide

CCCPC—Central Committee of the Communist Party of China (中国共产党中央委员会 or 中央委员会)

CPPCC—Chinese People's Political Consultative Congress (中国人民政治协商会议 or 政协)

DOE—US Department of Energy

DOD—US Department of Defense

EPA—US Environmental Protection Agency

GAQSIQ—General Administration of Quality Supervision, Inspection, and Quarantine (国家质量监督检验检疫总局 or 质检总局)

GHG—Greenhouse gases

IRS—US Internal Revenue Service

MEE—Ministry of Ecology and Environment, China (生态环境部)

MEP—Ministry of Environmental Protection, China (环境保护部 or 环保部)

MIIT—Ministry of Industry and Information Technology, China (工业和信息化部 or 工信部)

MOF—Ministry of Finance, China (财政部)

MOHURD—Ministry of Housing and Urban-Rural Development, China (住房和城乡建设部 or 住建部)

MOST—Ministry of Science and Technology, China (科学技术部 or 科技部)

MOT—Ministry of Transport, China (交通运输部 or 交通部)

MNR—Ministry of Natural Resources, China (自然资源部)

NASA—US National Aeronautic and Space Agency

NDRC—National Development and Reform Commission, China (国家发改和改革委员会 or 国家发改委)

NEA—National Energy Administration, China (国家能源局)

NHTSA—US National Highway Traffic Safety Administration

NOAA—US National Atmospheric and Oceanographic Administration

NPC—National People's Congress, China (全国人民代表大会or 全国人大)

NSC—National Security Council, the White House

SASAC—State-owned Assets Supervision and Administration Commission of the State Council, China (国务院国有资产监督管理委员会 or 国资委)

SAT—State Administration of Taxation, China (国家税务总局)

SFA—State Forestry Administration, China (国家林业局)

SOEs—State-owned enterprises (国有企业 or 国企)

USDA—US Department of Agriculture

1   Introduction

On November 11, 2014, President Xi Jinping (习近平) and President Barack Obama walked into a press conference in the Great Hall of the People adjacent to Tiananmen Square in Beijing and stunned the world with their Joint Announcement on Climate Change. For the first time ever, China announced that it would limit its emissions of heat-trapping carbon dioxide (CO_2) in absolute terms so that its emissions would peak around 2030, with "best efforts" to peak early. Relatedly, China also pledged to increase the share of nonfossil fuels in its primary energy consumption to around 20 percent by 2030. For its part, the United States committed to achieve an economy-wide target of reducing its greenhouse gas (GHG) emissions to 26–28 percent below its 2005 level by 2025 and to make "best efforts" to reduce its emissions by 28 percent (White House 2014).

When President Obama flew to Beijing, nobody predicted a major breakthrough in US-China relations, with one headline in the *Guardian* pessimistically predicting a widening in the "China-US gulf" due to the "marginalized" US president nearing the end of his administration after a devastating midterm election (Kaiman 2014). Numerous tensions had intensified that year, especially over cybersecurity, China's high- and new-technology enterprise (HNTE) policies, China's island-building program in the South China Sea, and human rights in Hong Kong.

Climate change was not considered a top-tier issue by the foreign policy establishment in either country,[1] and the United States and China had been deadlocked in the international climate negotiations for decades. The United States was the largest source of GHG emissions throughout the twentieth century and, as of 2014, had per capita emissions about three times higher than China's (WRI 2017). China's emissions had accelerated after the turn of the century on an exponential trajectory; China surpassed the United States in 2007 to become the largest annual emitter. These two major superpowers were engaged in what some observers called a "suicide pact," with neither willing to reduce emissions unilaterally (except on a voluntary basis) if the other did not make comparable commitments, dooming themselves and the world to substantial climate change in the future.

The United States and China are the titans[2] of the climate, together accounting for nearly 45 percent of global CO_2 emissions from fossil fuels. No other individual country comes anywhere close to having their impact on the pace and magnitude of climate change. China currently accounts for 30 percent of fossil fuel emissions and the United States for 14 percent (Olivier, Schwe, and Peters 2017). After these two countries, the entire European Union comes in third, India fourth, and Russia fifth on a national basis. The emissions of the next-five-largest emitters combined—Indonesia, Brazil, Japan, Canada, and Mexico—do not even equal China's total emissions in 2014. Cumulatively since 1850, the dawn of the Industrial Revolution, US emissions account for 27 percent of the global total and Chinese emissions for 11 percent (WRI 2014).

Defying Expectations Together

Given the sheer size of US and Chinese emissions, nearly half of the climate change problem was instantly addressed with the 2014 Joint Announcement. Adding in Europe, more than half of global emissions was controlled as of 2014. The following year, in December 2015, the world came together in Paris to negotiate a global accord, which was signed by 195 countries, a near-universal agreement. How and why did the United States and China defy expectations and agree to work together?

Europe was the original first mover in emissions reductions, having taken the previous international agreement—the Kyoto Protocol to the UN Framework Convention on Climate Change (UNFCCC)—seriously and reducing its emissions 8 percent below 1990 levels by 2012. In the original 1992 UNFCCC agreement, the United States agreed to "aim" to reduce emissions to 1990 levels, but as of 2014 it was 27 percent above those levels. China, as a developing country, had no binding obligations to reduce emissions in that agreement.

The United States did not adhere to most of its commitments under the UNFCCC, nor under its subsequent Kyoto Protocol, which the US Senate refused to ratify on the grounds that other major emitters, and specifically China and India, did not have similar obligations.[3] President Obama came into office with climate change as one of his highest priorities, but his first term was consumed by addressing the immediate economic recession and then passing the landmark Affordable Care Act, which was signed into law in 2010. During the second term of the Obama administration, however, President Obama announced a climate action plan that was intended to make good on the commitment he made at one of the UNFCCC's Conferences of the Parties held in Copenhagen in 2009—namely, a 17 percent reduction in greenhouse gases below 2005 levels by 2020.

By 2014, US emissions had apparently peaked and begun to decrease. Emissions reductions occurred in response to new policies, such as automotive fuel economy standards, and the market-based transition from coal to natural gas in US power generation due to the shale gas revolution that had resulted in plunging natural gas prices in the United States. Now that US emissions were on a downward trajectory, the United States finally had some moral standing in negotiations with other countries. President Obama was keen to contribute to the achievement of a global agreement on climate change in 2015, and was willing to take some risks to do so.

Meanwhile, China had adhered to the provisions of the UNFCCC and the subsequent Kyoto Protocol, although neither of those agreements required China, as a developing country, to reduce emissions unless it wanted to do so on a voluntary basis. Between 1992 and 2014, China's emissions skyrocketed 280 percent, surpassing those of the United States in 2007 and making China the largest overall emitter in the world on an aggregate but not per capita basis (World Bank 2017). After the UNFCCC was adopted in 1992, some senior Chinese officials thought that climate policy would constrain development, but a gradual shift in thinking occurred. Over time, multiple cobenefits to climate mitigation were recognized: structural economic reform to boost cleaner industries, conventional air pollution reductions, development of strategic renewable industries, international reputation, geopolitics, access to markets, and acquisition and development of advanced low-carbon technologies (Hu and Guan 2008).

As China's GHG emissions grew, so did conventional air pollution, reaching unprecedented levels after 2005. Between 2008 and 2015, air pollution was in the "unhealthy" or "very unhealthy" range in Beijing 40–60 percent of the time (BBC News 2015). In 2014, Beijing experienced a very bad episode of air pollution that quickly became known as the "airpocalypse." A documentary film by journalist Chai Jing (柴静), *Under the Dome* (苍穹之下), was released on February 28, 2015, that documented the increasing severity of urban air pollution around China through 2014. More than 300 million people viewed the film before it was taken offline seven days later. The government issued the first-ever "red alert" for Beijing in December 2015 when air pollution was so bad that schools had to be closed.

Another factor motivating the Chinese government was its long-standing desire to reform the Chinese economy away from the traditional development mode, which relied heavily on resource-intensive heavy industries. The Chinese government strategy for economic development from 2011 is aimed at reducing the carbon intensity of the economy, spurring innovation, upgrading manufacturing, being driven by innovation, shifting to a service-based economy, and ultimately achieving an "ecological civilization."[4] The Chinese government is still trying to alter the composition of the economy so that it relies more on technology-intensive industry and services, with emphasis on strategic industries, including alternative vehicles and clean energy. Such structural shifts greatly contribute to reducing both GHGs and urban air pollution. An international agreement requiring the Chinese government to reduce emissions would strengthen the hand of the central government in implementing economic structural reforms domestically. Naturally, energy-intensive and heavy industries and other vested interests have resisted these reforms, including some of the state-owned energy enterprises and some local governments, especially those with rich energy-resource endowments or where gross domestic product (GDP) and employment depend heavily on heavy industry.

The Chinese central government also had new global ambitions. During his trip to the United States in February 2012, President Xi declared, "We should … expand our shared interests and mutually beneficial cooperation, strive for new progress in building our cooperative partnership, and make it a new type of relationship between major countries in the 21st century." This concept, which did not originate with Xi Jinping (习近平), became known as the "new model of great power relations," a concept which was vague but intended to promote a mature relationship among major world powers, in which there could be disagreement on some issues but cooperation on others. Debate about this concept, and China's evolving role in the world, dates back at least to the late 1990s in China. The idea of a new type of major power relations was first articulated by former president Jiang Zemin (江泽民) in the context of Sino-Russian relations (Zeng 2016).

President Xi mentioned climate change in his 2012 speech in Washington, DC: "China and the United States should meet challenges together and share responsibilities in international affairs. This is what China-US cooperative partnership calls for and what the international community expects from us . . . We should increase cooperation on global issues such as climate change, counter-terrorism, cyber security, outer space security, energy and resources, public health, food security, and disaster prevention and mitigation" (Xi 2012).

Among others, former White House Chief of Staff John Podesta and colleagues took note of the concept and coauthored a paper in December 2013 with recommendations on how to operationalize the great power relations concept in US-China relations (Podesta et al. 2013). Just a couple of months later, President Obama invited Podesta to become his counselor in the White House, with climate change policy as a principal area of his responsibility. Meanwhile, the US State Department was beginning to consider the idea of a US-China agreement. Since the Copenhagen conference, Todd Stern, the US special envoy on climate change, had made a particular effort to improve his relationship with Chinese Minister Xie Zhenhua (解振华), the long-serving diplomat representing China in the international climate negotiations. Secretary Kerry was eager to accelerate progress for an agreement in Paris. On his first trip to China in April 2013, Kerry established a formal US-China Climate Change Working Group.

In November 2013, an academic outside of government sent a memorandum to Todd Stern and to John Holdren, science advisor to President Obama, recommending a bilateral agreement on climate change. Stern subsequently convened a group of US experts to discuss the risks, pros, and cons of such an agreement.[5] A few months later, Stern floated the idea to Minister Xie, and Secretary Kerry proposed to his counterpart, State Councilor Yang Jiechi (杨洁篪), that the two countries try to reach a climate agreement in time for the next presidential summit scheduled for November 2014. At the Strategic and Economic Dialogue (S&ED) that summer, the Chinese vice premier, Zhang Gaoli (张高丽), provided a formal positive response to the US proposal, and discussions commenced in earnest. The Chinese side suggested that the negotiations begin with a technical exchange to help each side understand the situation and prevailing understanding on the other side about what was possible, after which the political negotiations could begin in earnest.[6]

The negotiation objective was for each country to announce its Intended Nationally Determined Contribution (INDC)—in other words, its emission-reduction targets—at the presidential summit. The INDCs were supposed to be submitted to the UNFCCC well in advance of the Paris Conference of Parties, so the United States and China intended to kick-start the process with their joint announcement. An important principle for the negotiations was that each side's targets were nationally determined, even though President Xi and President Obama ultimately needed to be able to stand up next to each other and implicitly endorse the other's target. Negotiations concluded in time for the Presidential Summit in Beijing. Aside from the numerical targets for emissions peaking and reductions, and China's non-fossil target, the 2014 Joint Announcement also contained a provision to cooperate at the subnational level and, specifically, to hold a cities summit the following year.

Symbolically, the commitments by the United States and China were tremendously important because the United States was the largest advanced industrialized country emitter and China was the largest developing country emitter. If these two countries could agree to reduce their emissions, breaking through decades of disagreement, surely the rest of the world could come together in a global agreement.

One particularly sticky issue in the international negotiations was the principle of *common but differentiated responsibilities* (CBDR). This principle refers to the notion that, because industrialized countries had decades to develop without any restraints on emissions, they had the obligation to take on greater responsibility for emissions reductions and to contribute more money to help developing countries make the transition to cleaner economies. How to operationalize this principle had bedeviled the international negotiations for years. The US-China agreement embodied a solution to CBDR because it acknowledged the principle and operationalized it with different types of targets for the two countries. The United States had an absolute emissions reduction target, and China had a planned peak in emissions coupled with a target for nonfossil fuels as a percentage of primary energy. Minister Xie Zhenhua (解振华) repeatedly referred to the breakthrough on CBDR during the subsequent year of international negotiations leading up to Paris. Chinese experts also emphasized the breakthrough, noting that "coordination between nations at different stages of development and on different segments of the global chain of production is a difficult but necessary precondition to realizing a true low-carbon transition" (Zou et al. 2014).

As expected, the Joint Announcement in November 2014 surprised and delighted most in the global community. Several world leaders subsequently credited this agreement as being a key turning point that led to the Paris Agreement. Ban Ki-Moon commented, "The joint China-U.S. announcement signals the shared vision and seriousness with which the world's two largest economies are moving to a low-carbon future. It demonstrates strong leadership and momentum for a comprehensive global climate agreement in Paris" (Xinhua 2013). And in the analysis of the president of the World Bank, Jim Yong Kim, the Paris Agreement, "was a very specific constellation of events. This doesn't happen if the French aren't working on this for a year. It doesn't happen if Obama doesn't spend the time building relations with Xi. It doesn't happen without the Chinese-U.S. announcement. But it represents the biggest shift we have ever seen on this global crisis" (Davenport 2015). Christiana Figueres, executive secretary of the UNFCCC, reflected in April 2016 that the US-China climate agreement was an essential element, "without which I don't know if it would have been impossible but at least unlikely to get a Paris Agreement" (Figueres 2016).

The 2014 Joint Announcement also catalyzed other countries to negotiate joint agreements with the United States—most notably Mexico, which committed a few months later to a 25 percent reduction in GHG below business as usual by 2030 and an emissions peak of 2026. Many other countries announced their INDCs, and going into Paris more than one hundred countries had submitted INDCs (UNFCCC 2017a).

Perhaps most unexpectedly, the 2014 Joint Announcement created a new bright spot in US-China relations, transforming the concept of a "new type of major country relations (新型大国关系)" into a reality.[7] To American observers, China's decision to take on climate commitments was one of the first examples of China shouldering responsibility for a major global challenge. The announcement also demonstrated that the two countries could act in a mature and responsible manner, disagreeing vigorously about some issues but cooperating in the global interest on others. The announcement also represented the first time that an environmental issue reached the top tier in bilateral relations. To Chinese observers, the agreement also represented an opportunity to "invigorate" bilateral relations, especially in economic restructuring and international trade given how closely intertwined the two countries are economically (Zou 2014).

After the success of the 2014 announcement, both China and the United States wanted to maintain momentum and even accelerate it the following year at the next presidential summit in September 2015. Both countries were eager to resolve bilateral differences related to the Paris Agreement, begin implementation of the targets announced the previous year, and plan the first cities summit. At the 2015 S&ED, the two countries created a Domestic Policy Dialogue during which information could be shared about domestic implementation of the INDCs and lessons learned. The 2015 Joint Statement of Presidents Obama and Xi contained new announcements about domestic policies; most remarkable were President Xi's announcements that China would establish a national cap-and-trade program for certain sectors by 2017, along with a nebulous green dispatch policy. Meanwhile, the United States Environmental Protection Agency (EPA) had just announced its proposed regulations under the Clean Power Plan. China also surprised the world with a public commitment of RMB 20 billion (USD 3.1 billion dollars) for south-south climate finance (Xinhua 2015c).

The inaugural 2015 US-China Climate Leaders Summit was held in Los Angeles, California, just before the presidential summit. Dozens of mayors and governors from both countries made emissions commitments. On the Chinese side, a new Alliance for Peaking Pioneer Cities (APPC) was announced, and all of the APPC Chinese cities pledged to peak their emissions in advance of the national target of 2030. Beijing and Guangzhou committed to a remarkably early peak year of 2020. On the US side, eighteen cities, counties, and states made specific commitments. Los Angeles committed to a 45 percent reduction in greenhouse gases below 1990 levels by 2025, 60 percent by 2030, and 80 percent by 2050. Houston committed to a 42 percent reduction below 2007 levels by 2016, and Boston to a 25 percent reduction below 2005 levels by 2020 and 80 percent by 2050.[8]

Why This Book

Against this backdrop, the remainder of this book focuses on how the United States and China develop and implement climate change policy *domestically*.[9] The success or failure of the US-China climate agreements and the 2015 Paris Agreement will entirely depend on implementation of policies that contribute to achievement of the targets. Indeed, the Paris Agreement itself is inadequate to the task of preventing substantial climate change, and it will therefore be necessary to update and revise it in the coming years. US and Chinese leadership will undoubtedly be required again and again to create the conditions for new and improved global agreements throughout this century. Such leadership will only be accepted if the two countries honor the nationally determined contributions (NDCs) made under the Paris Agreement, which, importantly, rests on domestic policy implementation.

Although President Trump declared he would withdraw the United States from the Paris Agreement, future presidents are likely to adhere to the agreement—and subnationally, many governors and mayors have announced their intention to honor the US pledge anyway through an initiative called America's Pledge and its catchy hashtag, #WeAreStillIn.[10] Part of Trump's rejection of the Paris Agreement, much like his predecessor George W. Bush's rejection of the Kyoto Protocol, was based on willful ignorance of the steps China was taking to combat climate change. Trump perceives China primarily to be a traditional economic competitor, but apparently has not yet realized that China is decisively moving to develop and dominate new "strategic" industries, including the clean energy industry. It is possible that Trump may decide not to cede the clean energy industry (and its associated manufacturing jobs) to China, in which case he might indirectly support the US target for economic reasons.

Despite the Trump administration's reversal on the Paris Agreement, the Chinese government has determined to stay the course on both domestic and international climate policy. At the 2017 Nineteenth National Congress of the Chinese Communist Party (CCP), the first to take place after Trump's announcement, the CCP confirmed that the direction of climate policy will not change and that the most important climate policy measures will be continued into the future. Green (绿色) is regarded as one of five new visions for development (五大发展新理念), and promoting green development (推进绿色发展) is described elaborately in the Nineteenth National Congress report, which includes the following components: "Promotion of a sound economic structure that facilitates green, low-carbon, and circular development"; "create a market-based system for green technology innovation; develop green finance"; "promote a revolution in energy production and consumption"; and "build an energy sector that is clean, low-carbon, safe, and efficient" (Xinhua 2017b). Most of these plans are not new but it is important that they were reconfirmed in the face of Trump's stated intention to withdraw.

The bad news is that the high prioritization of climate change in China's overall policymaking agenda may decrease during the Trump era. Other issues like air pollution are more immediately pressing, and China's central government has lost the opportunity to use the climate change issue to foster a stronger relationship with the United States given the change in the US attitude toward the Paris Agreement. In official documents and files, the government will stay the course on climate policy, but in reality, the degree of importance and prioritization of climate change are likely to decrease significantly for some years.

The Nineteenth National Congress was also remarkable because it demonstrated the consolidation of Xi Jinping's (习近平) power. If Xi Jinping personally pays attention to the climate change issue, it is more likely that local cadres will implement domestic climate policies since they know it is one of his priorities. On the other hand, if Xi Jinping's interest wanes or if he becomes distracted with other issues, there is a possibility that climate policy could become a lower priority during the Xi Jinping era.

We decided to write a *comparative* book so that there would be two points of reference for readers. Climate policy in these countries is like a mosaic: it depends on your point of view. If people are more familiar with the US system, then they will be able to understand more easily the points of departure in the Chinese system, and vice versa. We intend to objectively help readers understand how and why the policy process differs in each country given the unique constraints and opportunities in each country. Although this book focuses on climate policy, it also provides a look at broader contemporary policy processes in both countries.

Perhaps because the policymaking systems are so different, and perhaps because of rising power rivalries, considerable distrust exists between the two countries (Lieberthal and Sandalow 2009), not only on issues like cybersecurity, intellectual property, foreign-direct investment, the South China Sea, and North Korea, but also on climate change. This distrust is fueled by persistent misunderstandings about each other that have led to the formation of prevailing myths that are wrong at best and slowing efforts to combat climate change more effectively at worst.

Two perfect and reciprocal examples of these myths are that (1) climate policy is a global conspiracy to keep China's development in check and, conversely, that (2) climate change is a hoax perpetuated by the Chinese government to undermine US competitiveness. Another foreign myth about China that must be examined is that if China's leaders just want to achieve something, they can merely issue an order and it will be obeyed (and therefore, if they fail to do so, they simply don't want to implement such a policy). One Chinese myth about the United States is that the United States is too democratic. In other words, there are too many voices, too many politicians, and therefore Washington lacks the ability to get anything done. Readers should be convinced by the end of this book that none of these myths are accurate portrayals of reality.

The central aim of this book is to help people understand the specific drivers of climate policies in both countries, and we group them into three categories: political, economic, and social. Seven factors are the most influential in differentiating climate policy outcomes in the United States and China: party politics, separation of powers, government hierarchy, individual leadership, economic structure and strategic industries, bureaucratic authorities, and the role of the media. The book will explain what each country has done in the climate change arena, why it has done so, and what it is likely to do in the future based on these factors.

In this book, we try to answer a number of questions that naturally arise about the differences in climate policy outcomes. The following are some of the main puzzles and questions that we seek to address:

• Why has climate policy been bottom-up in the United States and top-down in China?

• Why hasn't the United States chosen economically efficient, least-cost, market-based climate policies so far even though it preaches market-based approaches to policy internationally?

• Why is China beginning to embrace emissions trading even though it has virtually no experience with it at all, and why isn't the United States using it to address GHG emissions when it has successfully used this policy instrument already to address conventional air pollution?

• Why is the United States using a regulatory approach to climate policy even when it has historically claimed to prefer market-based approaches for environmental policy?

• Why and how did each country choose its national target and are they stringent enough?

• Why does it seem relatively easy for the United States enforce climate policies and more difficult for China to do so?

• How does individual leadership affect climate policy in both countries and how are leaders constrained differently?

• How do institutional and bureaucratic differences affect climate policy outcomes in each country?

• Why does China appear to welcome international agreements more than the United States does?

Main Arguments

First, we should not expect both countries to have the same approaches to policy because they have different histories, cultures, and institutions and are at different stages of economic development. They do not use have the same national circumstances, do not use the same criteria for evaluating potential policies, do not share exactly the same goals, and, thus, do not have the same policy outcomes.

Second, through better understanding of each other, we might be able to turn distrust into at minimum a cautious skepticism and perhaps, even better, a more realistic approach to cooperation. Empathy and understanding are the foundations for partnerships. Through better understanding about each other's differences and particular constraints, future negotiations between the two countries will be more productive because we will understand the fundamental interests of the other side. We need to be more knowledgeable, less suspicious, and more respectful about fundamental uncertainties, and we must learn from each other's experience.

The Chinese policy process exhibits strategic pragmatism compared with an American process characterized by deliberative incrementalism. Because the CCP is diffused throughout the government, it can take a long-term strategic approach to policy. Although the Chinese policy process can still be considered "fragmented authoritarianism" (Lieberthal and Oksenberg 1988), the consolidation of economic power since China's opening and reform now allows the Chinese government to concentrate financial resources behind its priorities in order to achieve its goals. The lack of term limits or elections of leaders means that the government can take a longer-term view, engage in strategic planning, and implement policies pragmatically and systematically. The Chinese people believe in a brighter society rooted in Confucian culture, in which an individual works to be moral and good, and the society works to be harmonious through everyone's efforts. The *China Dream*, a concept promoted by the CCP, embodies the idea of every Chinese citizen working toward a prosperous future through a process of "national rejuvenation" (Xi 2017).

The US policy process is characterized by deliberative incrementalism, with many steps forward and backward as new leaders are elected from different parties for varying periods of time. Election cycles come quickly, so politicians are focused on near-term, symbolic wins rather than pragmatic steps toward a long-term goal. Americans are also persistently optimistic about the future, embracing themes like America as a "shining city upon a hill" (Reagan), "hope" (Obama), and "making America great again" (Trump) due to their innate confidence in an individual's ability to achieve a better life. US government leaders look for incremental gains that, over time, metamorphose into major pieces of legislation like the Clean Air Act when the political stars align due to a slow development of national consensus. These concepts are elaborated on in chapter 7.

Finally, climate policy is a window into the broader policy processes at play in each country. We can use the insights gained from improved understanding of the climate policy process to understand more generally one another's policymaking structures, actors, processes, and approaches.

Roadmap of This Book

Chapter 2 provides a tour of the policy landscape in each country as of 2017. Each country's national circumstances are analyzed, sources of economic growth are identified, and existing policies are explained. Chapter 3 begins by assessing public perceptions of the other country's government and climate change itself, and then compares government structures, actors, processes, and approaches in a more systematic way, using climate policy for illustration. Chapters 4 and 5 provide two case studies of the policymaking process: chapter 4 is devoted to national target formation, or how countries set climate policy goals in the first place, and chapter 5 examines how those goals are implemented through specific policies and enforcement procedures. Chapter 6 synthesizes why policy outcomes differ in the two countries and emphasizes seven main distinctions. We conclude the book in the chapter 7 by revisiting the myths and questions identified above and more generally characterizing Chinese and American approaches to policymaking. The appendix provides a detailed compendium of the main national climate policies, organized by type, of both countries.

This book has been researched over the course of the last two decades as both authors have spent years of our lives in each other's countries, learning about their similarities and differences in policy, culture, government, business environment, education, and daily life. Both of us are academic researchers who have had direct experience in our governments. Kelly Sims Gallagher is a professor of energy and environmental policy at the Fletcher School, Tufts University, in the United States. From 2014 to 2015, she served as a senior policy advisor in the White House Office of Science and Technology Policy and in the Office of the Special Envoy for Climate Change in the US State Department. Xuan Xiaowei is a senior research fellow in the Department of Development Strategy and Regional Economy at the Development Research Center of the State Council of the People's Republic of China (the DRC). The DRC serves as the think tank for the central government and provides policy options and advice to the CPC Central Committee and the State Council and thus regularly interacts with senior Chinese leaders (DRC 2013).

The book is therefore based in part on our direct experience. It also relies heavily on primary sources—and especially on government documents. Existing scholarship, reports from research institutes, and media reports and articles are also essential sources. We completed a comprehensive inventory of national-level climate policies dating back to the year 2000, and this inventory is provided in the appendix to this book. By delving into two case studies in depth, we can trace the policymaking processes of both countries and gain insight into the messy process of how policymaking works in both countries.

In summary, we write this book primarily to demystify the two countries' approaches to public policy and to climate policy in particular. We have noticed how easy it is to slip into suspicion of the other side, how quickly mistrust can develop, and how relations can deteriorate rapidly over misconceptions. We identify some fundamental differences that explain why our two countries will inevitably have contrasting approaches to climate policy, but also identify the commonalities that can prove useful in understanding each other. As scholars, we endeavor to clarify the similarities and differences in policy process and wish to update the prevailing understanding of the policy process in both countries.

2   National Circumstances

Although the United States and China are the two largest overall emitters of greenhouse gases, their national circumstances could hardly be more different. Approaches to climate policy must begin with an understanding of the context in which policy is developed and implemented, and this chapter will begin by describing each country's energy endowments, levels of economic development, pressing current challenges, public attitudes, and experience of the phenomenon of climate change. The remainder of the chapter is devoted to sketching the landscape of the existing climate policies in both countries.

Energy Resources

The sources of energy supply greatly affect a country's emissions of conventional pollution and greenhouse gases. Of the fossil fuels, coal generally causes the most pollution and natural gas the least (with oil falling in between), but actual emissions depend greatly on the technology used for extraction, production, and consumption of the fuels. When used in power plants, coal is typically more than twice as carbon-intensive as natural gas, and coal-based technologies generally have higher emissions factors (see table 2.1). For both coal and natural gas–fueled power plants, it is technically possible to capture and sequester the resulting CO_2 emissions, although the cost of doing so remains prohibitive as of 2017. It is important to remember that if natural gas leaks during the extraction and production process, the overall GHG emissions increase dramatically because methane has a high global warming potential (GWP) in the atmosphere on ten- and twenty-year time scales (IPCC 2014). Oil typically is not used in power plants in either the United States or China and is mainly reserved

Table 2.1
Average direct emissions from power plant combustion

Type	Tons CO_2eq/MWh
Natural gas combined cycle	0.370
Coal pulverized combustion	0.744
Natural gas combined cycle with CCS	0.047
Coal pulverized combustion with CCS	0.121

Source: IPCC AR5, WGIII, annex II, table A.II.13, p. 1307, downloaded from https://www.ipcc.ch/pdf/assessment-report/ar5/wg3/ipcc_wg3_ar5_annex-ii.pdf.

for the transportation sector. Nuclear, hydroelectric power, solar, wind, and geothermal forms of energy are all relatively low-carbon and low conventional pollution sources of energy. None of them can be considered zero-carbon energy sources because the production of these technologies creates pollution.

Both China and the United States were blessed and cursed with gigantic natural energy resource endowments in the form of coal. Coal was a blessing because it fueled the Industrial Revolution, contributing to the economic development and poverty alleviation of both nations. It was a curse because coal is the most carbon-intensive fuel and the leading culprit in greenhouse gas emissions, conventional air pollution, acid rain, and health problems related to all of the above.

Just as China was launching its economic reforming and opening up process in 1987, the United States was estimated to hold larger quantities of recoverable coal reserves than China (see figure 2.1). As of 2016, however, China had boosted its recoverable coal reserves to be roughly equivalent to those of the United States. Even so, China utilized vast quantities of its coal to support its rapid industrialization strategy and as of 2016 only had approximately seventy-two years' worth of remaining coal at current rates of production. China's recoverable coal reserves account for 21 percent of the world total. The United States also relied primarily on coal for its industrialization during the late nineteenth and twentieth centuries, but it was able to diversity to other fuel sources after World War II. The United States still has 22 percent of the world's remaining coal reserves and a reserves-to-production ratio of 381 years.[1]

In terms of natural gas, both countries were also blessed with enormous but different natural gas endowments. Neither began to exploit them until

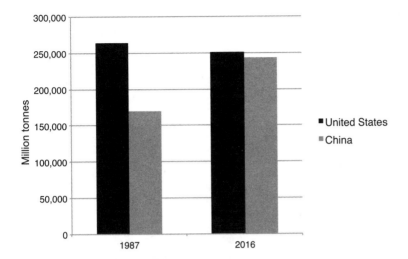

Figure 2.1
US and Chinese recoverable coal reserves.
Sources: For 1987 data, see Energy Watch Group, "Coal: Resources and Future Production," Paper No. 1/07, 2007, https://www.energiestiftung.ch/files/downloads/energiethemen-fossileenergien-kohle/ewg_report_coal_10-07-2007ms.pdf (accessed April 6, 2018). The Energy Watch Group's source was former editions of BP's "Statistical Review of World Energy." For 2016, the source is BP 2017.

the second half of the twentieth century. The United States began to shift to conventional natural gas in a serious way after 1950, and then was technologically and economically able to begin to exploit unconventional shale gas after the turn of the century. The abundance of relatively inexpensive domestic natural gas supply caused US electricity producers to dramatically shift from coal to gas for their fuel. This fuel switching has greatly helped to reduce CO_2 emissions in the United States since gas is a much less carbon-intensive fuel compared with coal. China had less conventional gas to begin with, but may have even more shale gas resources than the United States (even though they are still essentially uncharacterized and unexploited today). The geology in China for shale gas is much more complex and challenging than in the United States, and so the cost of shale gas production is so far too high to be profitable there except in some regions such as southwestern provinces Chongqing and Sichuan. Neither country dominates global reserves of natural gas the way they do in coal. Today, the United States holds just 5.2 percent of total gas reserves and China holds just 1.8 percent, even accounting for the new discoveries of shale gas.

Table 2.2
Estimates of renewable energy technical resource potentials (GW$_e$)

	China	United States
Hydro	400–700	153
Onshore wind	1,300–2,600	10,955
Offshore wind	200	4,224
Solar PV (utility)*	2,200	1,218
Solar PV (rooftop)	500	665

*Not including concentrating solar resource potential.
Sources: Data from IRENA REMap 2030 China, November 2014, and IRENA REMap 2030 USA, January 2015, Abu Dhabi, available at irena.org/remap.

Both countries are fortunate to enjoy excellent potential for renewable energy, including both wind and solar energy, as table 2.2 indicates. The technical potential for renewable energy the United States varies substantially by region but is large overall. Like in the United States, China's renewable resources vary substantially by region. China has excellent solar technical potential in the northern and western parts of the country, and China's best wind resources are also located in northern and western China, Inner Mongolia (内蒙), Xinjiang (新疆), Gansu (甘肃), and offshore (NEA 2017).

Energy Consumption

Now that we have established that there are plenty of energy resources of all kinds that are physically available, the next question is how much each type of energy is being utilized in each country. The short answer to that question is that China currently relies much more heavily on coal than does the United States, and the United States relies much more heavily on gas than does China. Both countries are racing to develop their renewable energy resources.

As of 2016, China relied on coal for 62 percent of its energy consumption, oil for 18 percent, gas for 6 percent, and nonfossil sources supply 14 percent (Wang 2017). Comparatively, in the United States, coal accounted for 18 percent of its total energy consumption, oil for 40 percent, natural gas for 25 percent, and nonfossil for 21 percent (EIA 2018a). This structural difference in starting points makes the climate change challenge incomparably more difficult for China than the United States. It is important to understand that China has made a tremendous effort to reduce coal as a fraction of its primary energy supply, with remarkable success. Almost twenty years ago, in 2000, coal accounted for 75 percent of primary energy supply, but China has diversified its fuel supply very quickly (IEA 1999, 56).

Another major structural difference is that industry accounts for a much larger fraction of energy consumption in China (49 percent) than in the United States (17 percent), meaning that China's economic growth is much more closely linked with its energy consumption and GHG emissions than is that of the United States (IEA 2017). Because many of China's manufactured goods are exported, it could be argued that importers are "outsourcing" CO_2 emissions through trade. In fact, China is by far the largest exporter of embodied CO_2 emissions in the world (Davis and Caldeira 2010). On the other hand, China's manufacturers provide employment and contribute to the national GDP, whereas other countries have suffered economically from the loss of manufacturing and its associated jobs.

Sources of Greenhouse Gas Emissions

The sources of GHG emissions vary dramatically across the two countries due to their different fuel sources and economic structures. Figure 2.2 illustrates how each country's sectors contribute to their overall GHG emissions. It is clear that the energy sector accounts for the majority of China's emissions due to its heavier reliance on coal. Within the energy sector for China, manufacturing industries account for 37 percent of total energy CO_2 emissions. The energy sector is also the largest source for the United States, although it is far smaller than China's in terms of emissions. Nonenergy industrial processes are also quite significant for China. In the transportation sector, however, the United States far exceeds China in terms of CO_2 emissions.

Economic Development

The economies of the United States and China are fundamentally different as of 2017. At the end of the twentieth century, the furious pace of economic growth in China contrasted greatly with slow growth in the United States. China was the world's manufacturing powerhouse, and the United

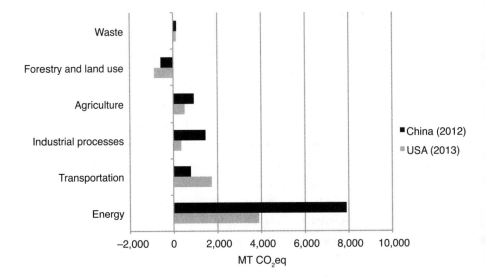

Figure 2.2
Sources of GHG Emissions (MT CO₂eq). Note that "industrial processes" do not include energy consumption by industry. "Energy" includes energy consumption across all sectors.
Sources: Government of the People's Republic of China 2016; US Department of State 2016. For the US data, all GHG emissions are included. For the China data, only CO_2 emissions are included for energy and transportation, but CO_2, CH_4, and N_2O are included for the other sectors.

States was experiencing a painful process of deindustrialization. Nearly one-fifth of the way into the twenty-first century, China's growth is beginning to slow, unemployment is rising, and China is trying to shift from an export-led manufacturing economy to one that is more balanced across different types of industries and that relies more heavily on domestic consumption. Although many Americans now think of China as a wealthy global economic challenger, most Chinese consider themselves to still be living in a developing economy, and indeed they are.

Per capita income in China is just $8,069, compared with $56,116 in the United States using conventional market exchange rates. Adjusting for purchasing power parity, China's per capita income rises to $14,451 (World Bank 2017). These national averages mask significant income inequalities: Chinese urban dwellers earn 2.7 times more than their rural counterparts (National Bureau of Statistics of China 2016, tables 6.6 and 6.11). The rural/

Table 2.3

Value added by industry as a percentage of GDP

	United States 2016	China 2014
Agriculture, forestry, fishing	1	10
Manufacturing	12	30
Mining	1	4
Construction	4	7
Wholesale and retail trade	12	10
Information	5	2
Finance, insurance, real estate	21	16
Professional and business services	12	4
Education and health care	9	5
Arts and recreation	4	<1
Government	13	4
Other	6	7

Sources: For the United States: Bureau of Economic Analysis, release date April 17, 2017, https://www.bea.gov/iTable/index_industry_gdpIndy.cfm. For China: *China Statistical Yearbook 2016* (National Bureau of Statistics of China 2016) available at http://www.stats.gov.cn/tjsj/ndsj/2016/indexeh.htm.

urban income divide is not nearly as wide in the United States, where rural residents earned $52,386 in 2016 compared with $54,296 for city dwellers. There is more income disparity among US states in rural areas: the top-earning state, Connecticut, has a rural per capita income of $93,382, and the poorest state, Mississippi, has a rural per capita income of $40,200 (US Census Bureau 2016).

The economic drivers by industry also vary dramatically across the two countries. As shown in table 2.3, finance, insurance, and real estate are the main drivers of the US economy, followed by manufacturing, trade, and services. In China, manufacturing still dominates the economy, accounting for 30 percent of value added, and finance, insurance, and real estate come in second at 16 percent.

Economic Growth Strategy

After 1978, the beginning of China's opening up to the rest of the world, the Chinese government pursued two economic development strategies:

(1) transition away from a planned economy and (2) adopt the East Asian development strategy used by Japan, Taiwan, and South Korea. For the second approach, the government took a heavy hand in industrial policy to promote particular sectors that could lead to exports, relying more heavily on foreign investment than did the other East Asian economies (Kroeber 2016). Looking ahead to the second quarter of the twenty-first century, the Chinese government strategy places greater emphasis on the role of the market, continued reform of state-owned enterprises (which is not new), and pursuit of the "circular economy" to reduce waste and inefficiency.

Even more specifically, China's Made in China 2025 (中国制造 2025) plan of 2015 aims to upgrade Chinese manufacturing by the year 2025, after which two more plans will be issued, culminating in the achievement of China as a "leading manufacturing power" by 2049, the one hundredth anniversary of the founding of the People's Republic of China (State Council 2015). The essence of the Made in China plan is to strengthen China's move into "advanced manufacturing," and away from traditional heavy industries. Ten key strategic industries are identified in the strategy, including information technology, robotics, energy-efficient and "new" energy industries, biological medicine, and transportation (Xinhua 2015).

The final element of China's economic strategy is a commitment to technological innovation. Former president Hu Jintao (胡锦涛) called this approach a "scientific development concept (科学发展观)," which was people-centered, comprehensive, coordinated, and focused on sustainable development (Fewsmith 2004). Xi Jinping (习近平) subsequently has emphasized a "dynamic, innovation-driven growth model" that is less resource-intensive and in a low-carbon development mode (Xinhua 2016). Xi Jinping has called for an "ecological civilization (生态文明)," which is understood to mean respecting and protecting nature, conserving resources, restoring the environment, recycling, and sustainable development (China Daily 2017).

The United States is experiencing a completely different economic transition, one of deindustrialization, loss of manufacturing, and much greater reliance on services, as is evident in table 2.3. President Trump made a campaign promise to bring manufacturing back to the United States, but it remains unclear whether he will be able to do so. US manufacturing employment fell from a peak of fourteen million in 2007 before the recession to a low of 11.5 million in 2010, and had risen to 12.4 million as of January 2017 (BLS 2018). The US government does not articulate and implement economic growth strategy as explicitly as does the Chinese government. In 2015, at the end of the Obama administration, the White House released a Strategy for American Innovation that is similar to Made in China 2025 in scope. In the US innovation strategy, three components are emphasized: (1) investment in research and development (R&D) and long-term economic growth, (2) identification of strategic areas such as precision medicine and advanced vehicles, and (3) making the federal government itself more innovative.

In his controversial book, *The Rise and Fall of American Growth*, Robert Gordon argues that innovation in the context of the second industrial revolution (1920–1970) drove US economic growth and improved human well-being in the United States at a pace that is unlikely ever to be matched again. He pessimistically forecasts lower economic growth and less rapid improvement in well-being in future decades based on the fact that US innovation since 1970 has been more confined to the information technology and communications sectors and is not as broad-based as it was in the first and second industrial revolutions. He also identifies other "head winds" in the US economy, including rising inequality, educational stagnation, declining labor force participation, and the fiscal demands of an aging population (Gordon 2016).

Climate Policy Landscape in the United States

After he was elected in 2016, President Donald Trump consistently directed his administration to make a concerted effort to roll back climate regulations issued under President Obama's leadership, and he also decided to withdraw from the Paris Agreement on Climate Change. EPA Administrator Scott Pruitt, a long-time opponent of environmental regulation as the attorney general from the oil and gas-producing state of Oklahoma, took on these tasks with vigor. In the first six months of the Trump administration, Administrator Pruitt filed a proposal of intent to undo or weaken the Clean Power Plan (CPP) and delayed a rule requiring fossil fuel companies to limit leaks of methane from oil and gas wells (Davenport 2017). Reportedly, he also initiated a formal program within the EPA to challenge mainstream climate science (Holden 2017). Whether or not the regulations targeted by Pruitt actually will be rescinded will be determined by the courts over many years. According to Harvard Law School professor Jody Freeman, "the Agency responsible for the rule … would need to go through a notice and comment process, which typically takes at least a year, often two, and sometimes longer. Challengers inevitably would sue over these actions, and the agency would need to defend them in litigation, mustering enough evidence to persuade a reviewing court that the agency is not being 'arbitrary or capricious.'"[2]

Under President Trump's predecessor, Barack Obama, climate policy shifted from a passive approach in prior administrations to a sharply focused plan to reduce GHGs. There was only one major new climate policy issued during the first term of the Obama administration; the rest were all components of the climate action plan released in 2013. The Obama administration's approach was to address all greenhouse gas emissions, not just CO_2, including emissions resulting from land-use change and forestry (LUCF).

US GHG emissions initially peaked in 2007 at 7,442 million metric tons of carbon dioxide equivalent (MMT CO_2eq). At the international climate change negotiations held in Copenhagen in 2009, President Obama committed the United States to reduce emissions "in the range of" 17 percent below 2005 levels by 2020. He further committed the United States to reduce to 26–28 percent below 2005 levels by 2025 in the international Paris Agreement. As of 2016, US GHG emissions, including net reductions from LUCF, had declined 10 percent from 2005 levels, putting the United States on track to achieve the Copenhagen target. Carbon dioxide accounts for the majority of US GHG emissions, and the main sources of CO_2 in the United States are electricity generation (36 percent) and transportation (36 percent). The largest industrial source of CO_2 emissions is the iron and steel sector, followed by cement and petrochemical production. The three largest sources of methane are, in order, enteric fermentation, natural gas systems, and landfills. The largest source by far of nitrous oxide (N_2O) is agricultural soil management, accounting for 77 percent of the total (EPA 2018).

The main sources of the reductions in GHG emissions in the United States since 2005 are in the electricity and transportation sectors. Electricity generation emissions declined 13 percent between 2005 and 2014, and transport emissions declined 8 percent. Slight reductions were achieved in the residential and industrial sectors, and a slight increase occurred in the commercial sector. The only major source of growth in emissions is from natural gas systems, which experienced a 40 percent increase in emissions during this period (EPA 2018).

These emissions trends are a product of market forces, human behavior, and public policy. Market forces contributed strongly to the decline in electricity sector emissions because the increased availability of cheap natural gas due to the US shale gas revolution caused many electricity producers to close older, carbon-intensive coal power plants and replace them with cleaner and more efficient natural gas and renewables-based power-generation facilities. In transportation, relatively cheap petroleum prices caused US consumers to drive their cars further and further each year, resulting in a 37 percent increase in vehicle miles traveled (VMT) between 1990 and 2014 (EPA 2018). This increased driving offset modest overall gains from tighter fuel-economy standards up until 2008, after which emissions from the transportation sector began to decline. As is evident in figure 2.3, the sales-weighted fuel economy of new US light-duty trucks rose dramatically after the Obama administration issued new fuel economy standards in 2012 that are supposed to rise to 54.5 miles per gallon by 2025 if the Trump administration does not alter them (White House 2012a).

In the electricity sector, CO_2 emissions from coal consumption declined 21 percent between 2005 and 2014, but CO_2 emissions from natural gas increased 39 percent. Renewable electricity generation from utilities and distributed generation grew 46 percent between 2006 and 2015 (EIA 2016a).

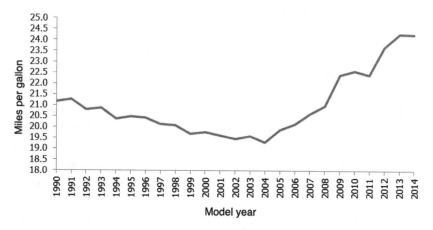

Figure 2.3
Sales-weighted fuel economy (in MPG) of new US light-duty vehicles.

The reduction in coal generation, growth in natural gas–fired plants, and growth in renewable electricity resulted in a net reduction in GHG emissions from the electric sector of 15 percent between 2005 and 2015 (EPA 2018).

The US Congress has never passed any form of comprehensive climate legislation. Congress did, however, support a number of crucial fiscal incentives in support of renewable and low-carbon energy sources that stimulated the private market for renewable energy during the Obama administration. The most notable incentives were the loan guarantee program administered by the DOE and the production and investment tax credits for renewable energy. The loan guarantee program was established in the Energy Policy Act of 2005 and considerably strengthened in the American Reinvestment and Recovery Act of 2009. The Loan Program Office had issued USD 32 billion in loans and loan guarantees for clean energy projects in the United States as of June 2016 (DOE n.d.-b). The production tax credit for renewables was extended in 2015 as part of the Consolidated Appropriations Act through 2019 (DOE n.d.-d). This same act extended investment tax credits for renewables through 2022 (DOE n.d.-a). Congress also supported small increases in energy-technology research, development, and demonstration throughout the Obama administration (Gallagher and Anadon 2017).

Although President Obama repeatedly expressed his preference for a market-based approach that relied either on a cap-and-trade program or carbon taxes, the US Congress refused to pass climate legislation. This reality forced the Obama administration to use its regulatory authorities under existing climate legislation to reduce GHG emissions where it could. The most significant regulatory actions that are in force and have already survived challenges in the courts relate to the transportation sector. Since 2009, using authorities derived from the Clean Air Act, the US EPA and National Highway Traffic Safety Administration issued two new phases of standards for light-duty passenger vehicles that will ultimately require new fuel economy to achieve 54.5 mpg and lower CO_2 emissions to 163 grams per mile in 2025 (US Department of State 2016). Similarly, the EPA and DOT issued the first-ever GHG and fuel economy performance standards for heavy-duty vehicles in 2011.

In the electricity sector, the signature initiative of the Obama administration was the Clean Power Plan, which would have reduced CO_2 emissions from power plants under the Clean Air Act 32 percent from 2005 levels by

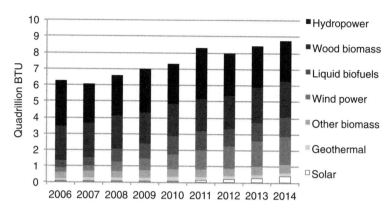

Figure 2.4
US renewable energy supply, 2006–2015. *Source:* EIA 2016b.

2030. The EPA issued a final rule for the CCP in August 2015, but the rule was immediately challenged in the courts. As of the writing of this book, the Supreme Court has not yet ruled on whether the regulation can enter into force. Meanwhile, the Obama administration committed to permit 20 GW of renewable energy on publicly owned lands through the US Department of the Interior (DOI). As of 2016, more than 10 GW of renewable energy were permitted. USDA also accelerated renewable energy deployment through its Rural Energy for America Program, which had awarded USD 789 million for 10,700 projects to install renewable energy systems or make energy-efficiency upgrades in rural small businesses, farms, and ranches (US Department of State 2016). Figure 2.4 depicts the growth in renewable energy supply between 2006 and 2015.

The US Department of Energy has existing authority under the Energy Policy Act to promulgate efficiency standards for appliances and industrial equipment, and it updated or issued thirty-four new efficiency standards between 2009 and 2015. The federal government does not have regulatory authority for building codes, but the Obama administration did create a nonbinding Better Buildings Challenge to promote a 20 percent improvement in energy efficiency in buildings by 2020. All of the efficiency standards finalized by the Obama administration have remained in force during the Trump administration as of early 2018. One of the standards that had not been finalized at the end of the Obama era was for ceiling fan efficiency, so nine states and New York City sued the DOE, after which the

DOE published a notice that it would finalize the ceiling fan standards and set compliance dates (HLS 2018).

Although CO_2 is the dominant GHG in the United States, the non-CO_2 gases are crucial to target with climate policies because their emissions are substantial and because some can be reduced quickly with near-term impact on the climate. To reduce emissions of hydrofluorocarbons (HFCs), the EPA finalized a rule in 2015 to prohibit use of some of the most GHG-intensive HFCs and to expand the list of acceptable alternatives under its Significant New Alternatives Policy (SNAP) program. The EPA also proposed a rule in 2015 to reduce methane emissions from the oil and gas sector and issued two proposals to reduce emissions from landfills. These proposed rules will need to be finalized and survive court challenges to become operational.

Given the intransigence of the US Congress, state and local governments have established most of the new climate policies in the United States since 2005. Thirty-seven states have renewable portfolio standards mandating a certain percentage of renewable energy production by certain dates. Two major subnational emissions-trading programs exist. The first is the Regional Greenhouse Gas Initiative (RGGI) of the northeastern states to reduce power-sector emissions, which revised its program to reduce its cap on CO_2 emissions by 2.5 percent per year through 2020 to achieve a level approximately 50 percent below 2005 levels. Second, in the State of California, the Global Warming Solutions Act requires GHG reductions to 1990 levels by 2020, including through a cap-and-trade program (US Department of State 2016).

In table 2.4, the most important federal and state climate policies discussed thus far are listed, along with the US government's estimated mitigation impact in 2020. As the table shows, the HFC regulation, vehicle performance standards, Clean Power Plan, and efficiency standards are estimated to have the largest impacts on CO_2 emissions by 2020. Without the Clean Power Plan or compensatory policy at the state level, it will be very difficult for the United States to meet its 2025 target. The appendix provides a complete chronology of climate policymaking in the United States.

Climate Policy Landscape in China

To explore the Chinese climate policy landscape, the best place to start is by examining the evolution of the policies over time given China's strong

Table 2.4

US federal climate policies and their estimated mitigation impact in 2020

Policy	Status of implementation	Type of policy instrument	Estimate of annual mitigation impact in 2020 (MT CO_2eq)
EPA SNAP program for ozone-depleting chemicals	Implemented	Regulatory	317
Light-duty vehicle CO_2 performance standards	Implemented	Regulatory	236
Energy-efficiency standards for appliances, equipment, and lighting	Implemented	Regulatory	216
ENERGY STAR labeling programs	Implemented	Voluntary	136
Investment and production tax credits for renewable energy*	Implemented	Fiscal	39–101
Clean Power Plan**	Rule finalized, being revised or repealed as of 2017	Regulatory	69
New source performance standards for volatile organic compounds for the oil and gas sector	Implemented	Regulatory	48
Heavy-duty vehicle CO_2 performance standards	Implemented	Regulatory	38
DOE loan guarantee program for clean energy	Implemented	Fiscal	14

*Estimates for production tax credits only and reported as an annual estimate calculated from a cumulative estimate between 2016 and 2030 from NREL 2016.
**With rate-based approach in 2020, rising to 415,000 kt of CO_2 in 2030, according to https://www.epa.gov/sites/production/files/2015-08/documents/cpp-final-rule-ria.pdf.
Sources: US State Department 2016, NREL 2016, and EPA Archive 2017.

tradition of policymaking through its five-year plans (五年规划). Because of the bureaucratic ownership of certain types of policies, we will also examine the Chinese climate policy landscape by ministry, agency, and department.

The most important policies are those targets (or indicators) specifically set by the central government in response to climate change, such as carbon intensity, energy intensity, total energy consumption, overall coal consumption caps, and the share of renewables in the overall energy mix. Some, but not all, of these targets are set in the context of the five-year plan. One target that first appeared in the Joint Announcement was the goal to peak emissions around 2030. The target and assessment process is mainly undertaken by government ministries and departments.

Responsibilities for achieving these targets are further divided at each level of government. Then, the higher level of government assesses the performance of the lower level; for example, the central government sets targets and assesses the performance of provincial governments. In some cases, the central government will supervise local governments directly, such as at the end of the eleventh five-year plan, when the national energy-intensity target proved difficult to achieve and the central government sent supervisory teams to assist local governments with their compliance (Legal Daily 2006). This target-responsibility system can directly influence the behavior of governments at all levels.

Since 2000, the Chinese central government has promulgated nearly one hundred significant climate change policies (see the appendix for a complete list). Most of these major policies have been revised and updated over time, usually in coordination with China's five-year plans. Many dozens of narrower policies also exist at the central government level, and as in the United States the provincial and local governments have many of their own subnational climate policies as well. Of the nearly one hundred policies, a subset of the most important ones is highlighted in table 2.5.

Many of China's climate change policies were initiated during the late 1990s and early 2000s, perhaps inspired by the 1992 Earth Summit and the new international agreement on climate change. The UN Framework Convention on Climate Change was adopted in 1992, and China ratified it early in 1993 (UNFCCC 2017b). China's climate policies have always been considered in a "cobenefits" context because most policies that have been pursued also provide benefits for reducing conventional air pollution, improving energy security, and boosting economic growth. In fact,

Table 2.5

Chinese central government climate policies and their average annual estimated mitigation impact between 2006 and 2015

Policy	Status of implementation	Type of policy instrument	Estimate of annual mitigation impact, MT CO$_2$*
Achieving 15 percent nonfossil energy by 2020 and 20 percent by 2030	Implemented	Regulatory	194
Raising share of natural gas to 10 percent by 2020	Implemented	Regulatory	58
Energy-efficiency labeling on end-use products	Implemented	Regulatory	32
Energy conservation in industrial sector**	Implemented	Regulatory	423
Tax incentives for renewable electricity	Implemented	Regulatory	63
Light- and heavy-duty vehicle energy-efficiency standards	Implemented	Regulatory	n/a
Purchase incentive subsides for new-energy vehicles	Implemented	Fiscal	n/a
Energy-efficiency standards for appliances, industrial equipment, and lighting	Implemented	Regulatory	n/a
Investment and production tax credits for renewable energy	Implemented	Fiscal	n/a
Phasing out small and outdated coal-fired power plants	Implemented	Regulatory, Industrial	n/a
National emissions-trading program	In development, as of 2017	Market-based	n/a
National coal-consumption cap	Implemented	Regulatory	n/a
Resource taxes for natural gas, oil, and coal	Implemented	Fiscal	n/a
Industrial upgrading and transformation	Implemented	Industrial	n/a
Top-1,000 and Top-10,000 Energy-Consuming Enterprises Program[†]	Implemented, expired 2016	Regulatory	76
Green credit policy	Implemented	Regulatory	n/a
Green bond policy	Implemented	Fiscal	n/a

*Based on an average of 2006–2015.

**Average calculated from 2011–2014.

***Average calculated from 2011–2015

†May include some double-counting with the "energy conservation in industrial sector" line above.

Sources: Some policies derived from the appendix. Other policies and all emission estimates from Government of the People's Republic of China (2016), "First Biannual Update Report on Climate Change," submitted to the UNFCCC, accessed from http://unfccc.int/national_reports/non-annex_i_natcom/reporting_on_climate_change/items/8722.php on 7/14/17.

the primary motivator for many energy policies is often one of these other objectives, but climate cobenefits are welcomed.

In the early years, the government's focus was on energy efficiency, motivated by a desire to reduce waste, promote economic development, and enhance energy security. China's design standards for energy efficiency in buildings were first issued in 1986 (later updated in 1995, 2010, and 2015). China's first policy on promoting the efficient development of coal-fired power generation was announced in 1999 and then was repeatedly revised through 2016. Similarly, China's first capacity-elimination program was ordered in 1999, the objective of which was to eliminate "backwards" production in various major industries.

In China's tenth five-year plan (2001–2005), the Chinese government concentrated heavily on economic growth. An annual growth rate target was set at 7 percent, but in reality it averaged 9.5 percent. Relatedly, the government aimed to increase government spending on R&D to more than 1.5 percent of GDP, thereby "speeding up technological progress" (China. org.cn n.d.). China's Ministry of Science and Technology (MOST, 科技部) established major energy RD&D programs on "new-energy" vehicles (fuel cells, electric, and hybrid-electric cars) and advanced coal (including both pre- and postcombustion CO_2-capture technologies) in the context of the 863 "high-tech" national R&D program.

From a climate point of view, probably the most far-reaching policy in the tenth five-year plan period was China's establishment of a concession incentive for renewable energy, coupled with preferential tax policies for renewable energy. Relatedly, China passed its Renewable Energy Law in 2005 and amended it in 2009. A concession program for wind power was introduced in 2003 via an NDRC directive, and a feed-in tariff (FIT) was established in 2009. A solar FIT soon followed. These two FITs were updated regularly, and the latest update, a notice on adjustments to the feed-in tariffs for onshore wind and photovoltaic (PV) power, was released by the NDRC in December 2016. China introduced the Golden Sun program in 2009 to provide support to its domestic solar industry after antidumping moves by the United States and Germany. The Chinese government initially provided an upfront 50 percent subsidy for grid-connected PV projects and incentivized off-grid solar power in rural areas to promote electrification of villages through a 70 percent subsidy (IEA 2017a). These generous subsidies were scaled back in 2012 as the cost of solar PV technology fell.

The State Council announced China's first comprehensive national climate guidelines in 2007 (中国应对气候变化国家方案). The National Climate Change Programme outlined objectives, basic principles, key areas of actions, and initial policies and measures to address climate change. Although guidelines are not formally binding, they serve the purpose of sending a clear signal to all levels of government about the importance of a particular issue—in this case, climate change. More detailed policies usually follow and are issued by the relevant ministries. This guideline probably was released in anticipation of the planned conclusion of the international climate negotiations in Copenhagen, scheduled to end in 2009 (these negotiations failed, and the Paris Agreement was not reached until 2015).

During China's eleventh five-year plan (2006–2010), much attention was devoted to improving energy efficiency, which has the cobenefit of reducing GHG emissions (Qi and Li 2011), because China's emission intensity began to increase again during the tenth five-year period due to increased output from heavy industry and manufacturing and due to weak regulations on energy intensity (Zhang and Huang 2017). During this five-year plan, the Energy Conservation Law was revised in 2007, and the State Council issued a Comprehensive Work Plan on Energy Conservation and Emissions Reductions. The latter imposed mandatory energy-intensity-reduction targets for local governments, and an overall national target was established. This target was updated in 2011 and 2016 in the twelfth (2011–2015) and thirteenth (2016–2020) five-year plans, respectively. The NDRC (国家发改委) and GAQSIQ (国家质检总局) also issued the first energy-efficiency labels in 2004 (revised in 2016). China's first fuel-efficiency standards for passenger vehicles were issued in 2004. In 2006, China's Top-1,000 Energy-Consuming Enterprises Program was initiated (NDRC 2006). This program is a mandatory energy-conservation target-setting policy for large industrial energy users. In exchange for meeting these mandatory targets, firms were provided with subsidies to upgrade to more energy-efficient equipment. In 2011, this program was expanded to include the top ten thousand enterprises.

Another major focus of the eleventh five-year plan period was the Circular Economy Promotion Law (循环经济促进法) and its accompanying guidelines issued by the State Council, the NDRC, and MIIT. In 2004, the State Council issued "Suggestions on the Developing of the Circular Economy," and these suggestions were turned into guidelines and action plans over time (State Council 2005a). The main goals of the circular economy concept were to reduce waste and promote the reuse of resources and materials.

The twelfth five-year plan (2011–2015) was an active period in Chinese national climate policymaking, initiated by President Hu Jintao (胡锦涛) and Premier Wen Jiabao (温家宝) and later evolving with Xi Jinping's rise to the presidency in 2013 after becoming the general secretary of the CCP in 2012. During this period, both the 2014 US-China Joint Announcement on Climate Change and the negotiation of the Paris Agreement on Climate Change took place. A new National Plan on Climate Change (2014–2020) was released by the NDRC, which updated and confirmed the targets established under President Hu and Premier Wen. The target of reducing CO_2 intensity by 40–45 percent per unit of GDP below 2005 levels by 2020 and a target of increasing the percentage of nonfossil fuels in primary energy consumption to 15 percent by 2020 were both included in the new national plan. This was the first five-year plan that included a national strategy for climate adaptation.

During the eleventh and twelfth five-year plans, a major effort was initiated to promote electric vehicles (EVs) based on the theory that China's manufacturers could leapfrog over the internal combustion engine and move directly to electric vehicles. To Chinese policymakers, the benefits of achieving such a leapfrog are substantial because China would not need to import as much oil (having become a net oil importer in 1993), China could rely less on foreign auto manufacturers for technological know-how (Gallagher 2006), and urban air pollution could be reduced. From a climate point of view, however, electric vehicles that are fueled by coal-fired power plants actually will lead to a net *increase* in emissions (Ji et al. 2012), so for EVs to be a good climate-mitigation strategy in China, the electricity system would need to move away from coal. In 2009, a demonstration program of energy-efficient and alternative-energy vehicles known as the Ten Cities, Thousand Vehicles program was released that was intended to stimulate EV development and deployment through generous subsidies. The program was subsequently expanded to eighty-eight cities. In 2012, the State Council issued an Energy-Saving and New-Energy Vehicle Industry Development Plan which targeted the production of five hundred thousand battery electric vehicles (BEVs) and plug-in hybrid electric vehicles (PHEVs) by 2015, with the production capacity to grow to two million units by 2020. All of these EV plans and guidelines were supplemented by the provision of subsidies beginning in 2009 (updated in 2013, 2015, and 2016). Coupled with these intensive efforts to leapfrog to electric vehicles, the Chinese government imposed more stringent fuel-economy standards on passenger vehicles in 2015 and introduced its first fuel-economy standards for heavy-duty trucks in 2015 as well.

During the twelfth five-year plan, the central government launched seven pilot cap-and-trade programs for greenhouse gases in different regions in order to learn and plan for a national emissions-trading program. The regions selected were Beijing (北京), Chongqing (重庆), Shanghai (上海), Tianjin (天津), Guangdong (广东), Hubei (湖北), and Shenzhen (深圳). In the context of the second Obama-Xi announcement in September 2015, the Chinese government committed to initiating a national cap-and-trade program for carbon dioxide in 2017, and this policy was announced for the power sector at the end of 2017. Also, China's Ministry of Finance completed a resource tax reform between 2011 and 2014, via which new taxes were levied on crude oil, natural gas, and coal based on the retail price rather than production.

Most recently, the thirteenth five-year plan (2016–2020) continues and deepens many of the policies already initiated during earlier periods. A new plan for energy development was released by the NDRC in 2016, which set an overall cap on primary energy consumption at five billion tons of coal equivalent by 2020. For the first time, a national coal cap was also set during this period in a new NEA policy according to which annual coal consumption should not exceed 4.1 billion tons before 2020. In a clear sign of the intention to shift to cleaner fuels, new targets were also clarified for nonfossil fuels, with the goal of achieving at least 15 percent of primary energy supply from nonfossil fuels by 2020. To achieve this ambitious goal, installed nuclear capacity should reach 58 GW by 2020, installed capacity of hydropower should reach 340 GW, wind 210 GW, and solar 110 GW.

For the second time, the State Council also released a specific Work Plan for Greenhouse Gas Emission Control for the thirteenth five-year plan ("十三五"控制温室气体排放工作方案). Along with setting a new target for a reduction in carbon intensity, the NDRC also released a work plan for the pilot construction of climate-resilient cities in 2016.

Although this book is primarily focused on national-level governance, it is crucial to understand that most of the implementation of central-government policies occurs at the subnational level. In addition, some provinces, cities, counties, and towns pursue their own climate policies (although they vary tremendously in their approaches and stringency), and usually local governments will obtain approval from higher-level governments before implementing any of their own new initiatives and policies.

3 Comparing Policymaking Structures, Actors, Processes, and Approaches

How does the policy process work in each country? In this book, we think of public policy as the decisions of government, whether they are formalized in laws, statutes, executive decisions, or commonly understood norms that influence human behavior (Weible 2014, 5). As a comparative analysis, we are most interested in how and why the policy development and implementation process varies in ways that produce different outcomes in the two countries.

Government decision-making is a source of mystery and confusion to observers on each side of the Pacific Ocean, given the vast differences in the historical development of governance systems in the two countries. Each country's approach to policy is uniquely different, though a remarkable number of similarities exist as well, which are illuminated in this chapter.

In the United States, contemporary understanding about the Chinese policy process emerged slowly after the reestablishment of diplomatic relations in 1979, initially impeded by the closed system in China and the relatively few number of scholars who were able to penetrate the Chinese government bureaucracies to conduct interviews. Gradually, access improved and some foreigners even began working in a collaborative fashion with Chinese academics and government officials on policy analysis. Even so, US scholarship on Chinese policy process waned after the seminal works published by Kenneth Lieberthal and Michel Oksenberg (1988) and others due to stronger interest among most American political scientists in understanding China's elite politics and China's communist party. Within China, political science was not encouraged as an academic discipline after the founding of the People's Republic of China in 1949, and this field of study essentially disappeared during the Cultural Revolution (Lieberthal, Li, and Yu 2014). Chinese study of both Chinese and American policy processes thus also is relatively recent but has grown in recent years. One milestone was the establishment of the Institute of American Studies under the Chinese Academy of Social Sciences (中国社会科学院美国研究所) in 1981, which is engaged in studying the politics, economy, society, and policies of the United States.

This chapter updates and clarifies the understanding about the policy process in the United States and China developed by scholars and observers during the past thirty years and fills knowledge gaps, particularly on the policy process in China. Before we dive into the current understanding of the policy process, it is worth first examining public perceptions of the other country's government.

Public Perceptions of the Other Country's Government and Climate Change

Public perceptions are shaped by many factors—and increasingly by social media. In China, the media are fully controlled by the state, and the most commonly used apps used in the United States, such as Facebook, are prohibited in China. On the other hand, many social media platforms like WeChat flourish in China, even if they are monitored. It is thus reasonable to conclude that the Chinese government shapes Chinese public perceptions to a much greater extent than does the US government. In fact, the widespread use of social media in the United States has allowed climate skeptics and self-interested parties to sow doubt and confusion about climate change through the promotion of "alternative facts," no matter how incorrect they are scientifically.

The American public tends to make a number of erroneous or questionable assumptions about China. A commonly held misperception, for example, is that because China has a single, authoritarian ruling party, the central government can implement any policy that it chooses. Politics, according to this view, does not interfere with the rational policymaking process. Another is that because China is a communist country, it does not have a market economy. Yet almost half of American citizens believe China will become the world's top superpower, or that it already is, supplanting the United States in this role (Wike 2016).

Conversely, many ordinary Chinese believe that the US president has more power to create policy than she or he actually has given that Congress is the only entity with constitutional authority to pass laws or authorize expenditures in the United States. On the other hand, most Chinese do not understand what types of executive authority are granted to the president. A large number of Chinese citizens believe that the United States is trying to limit China's power (Wike, Stokes, and Poushter 2015).

Climate change in particular is a hot-button topic that provokes suspicion on both sides. Some Chinese experts and senior government officials, while not questioning the existence of climate change itself, believe that climate policy is one particular means of holding back China's development (Qi and Chen 2017). In the United States, future President Trump himself asserted in a 2012 tweet that "the concept of global warming was created by and for the Chinese in order to make U.S. manufacturing non-competitive" (Trump 2012). This tweet was retweeted 105,211 times and received 67,509 likes.[1]

Public perceptions about the climate change issue vary substantially across the two countries. According to a study by Tien Ming Lee and coauthors (2015), more than 75 percent of the population in the United States is aware of climate change, but only slightly *more than half* view it as a threat. In China, slightly more than half of the population is aware of climate change, but *less than half* view it as a threat. The strongest predictors of Chinese risk perceptions are the belief that global warming is human-caused and dissatisfaction with local air quality. By contrast, American citizens who perceive climate change risks to be high believe that it is caused by human-caused pollution, believe that local temperatures are rising, and generally believe that the government should preserve the natural environment. A survey of college students in the United States and China found that a substantially larger proportion of Chinese students accepted the scientific consensus on anthropogenic climate change (Jamelske et al. 2015). In contrast, a higher proportion of US students reported being unconcerned about climate change compared with Chinese students.

Very different from the situation in China, there is a widening divide between the two main political parties in the United States on this issue. Democrats are far more likely to be concerned about climate change than are Republicans. This issue is discussed more extensively later in this chapter and in chapter 6, but the difference has arisen because Republicans dislike regulation and have become the main party supported by the fossil fuel industry. After the major environmental laws such as the Clean Air Act and Clean Water Act of the 1970s were passed, many industries believed there had been a regulatory overreach. When Republican Ronald Reagan was elected president in 1981, he embraced this thesis and worked to roll back many regulations (Yang 2017).

On the whole, however, despite the population having less awareness of climate change, Chinese people appear to be much more willing to pay to address the problem than American people, with 68 percent of Chinese surveyed supporting paying 1 percent of GDP to address climate change, compared with only 48 percent of Americans in a poll in 2009. In terms of their sense of responsibility to address climate change, Chinese citizens ranked second only to Bangladesh, with 98 percent believing that their country bears responsibility. In contrast, only 82 percent of Americans share this belief. Seventy-seven percent of Chinese did not think their country was doing enough in 2009, compared with only 58 percent in the United States (World Bank 2009). The 2015 survey found that Chinese university students were more supportive of joining an international agreement to address climate change than American students were (Jamelske et al. 2015). The same authors conducted a 2017 survey of adults in the two countries and found that a supermajority of adults in both countries were supportive of an international agreement on climate change, but that Chinese respondents were even more supportive than Americans. All else being equal, the researchers also found that more exposure to media content on climate change increased support for a treaty and that political affiliation strongly influenced support for a treaty in the United States (Jamelske et al. 2017).

Still, these public perception figures should be treated with great caution because there are huge differences between urban and rural Chinese, coastal and inland Chinese, and developed and underdeveloped regions in China; similarly, a wide gulf exists in the United States between "red" (Republican leaning) and "blue" (Democratic leaning) states, urban and rural regions, and among the Northeast, Southeast, Midwest, West Coast, Rocky Mountain region, and Southwest.

The rest of this chapter examines American and Chinese approaches to the policymaking process, with an emphasis, of course, on climate change policy. This chapter, and indeed the rest of the book, relies on the existing scholarly literature; published laws, regulations, and policies; personal experience serving in government and interacting with the policy processes in both countries; and interviews with current and former government officials, experts, business leadership, and individuals from nongovernmental organizations (NGOs). All interviews were conducted on a nonattribution basis to allow people to speak freely.

To clarify the similarities and differences between the policymaking processes of the two countries, we developed a comparative framework, which is presented in the next few sections of this chapter. This framework compares the structures, actors, and processes of the United States and China. We conclude the chapter by reflecting on the advantages and disadvantages of the two policymaking systems.

Comparing Policymaking Structures

For countries with such different political systems, one would naturally expect that the policy processes would be unrecognizable to each other. In some ways, they are. One-party rule allows the Chinese government to avoid the nearly complete legislative gridlock that has become the rule, not the exception, in the federal policy process in the United States since the 1990s. The US judicial system allows the federal government to implement policies at every level of governance, whether in a powerful state like California, a rural jurisdiction like Jefferson County in Colorado, or a small town like Belmont, Massachusetts, yet the Chinese central level of government struggles to implement and enforce national policies that have been approved by the Central Committee of the Party or the State Council. The central government is responsible for all the key decisions, but local governments take the lead on policy implementation.

Yet for all that is different between the two countries, much is the same. Individual people serve as policy entrepreneurs, utilizing their personal networks to advance ideas and shape policies or to block policies they dislike. The bureaucracies in both countries are powerful and competitive with each other, vying for the attention of senior leaders. Government officials must negotiate with each other to advance policy agendas. Interest groups and companies try to maximize their gains while minimizing losses as each policy gains momentum. In this chapter, we tease out the main similarities and differences between China and the United States.

Table 3.1 provides a comparative snapshot of the different structures in the US and Chinese policymaking systems. The United States is a democratic state with elected leaders, and China has a one-party state with

Table 3.1

Comparative structures of the US and Chinese policymaking systems

United States	China
Democratic state with elected leaders	One-party state with selected leaders
Republican and Democratic parties dominate, other parties allowed (e.g., Green Party)	Chinese Communist Party (CCP); other parties exist under supervision of the CCP and accept CCP leadership
Federal, state, and local government departments and agencies	Central (中央), provincial (省级), prefecture (地级), county (县级), and township (乡镇级) government ministries and bureaus*
Separation of power with equal executive, legislative, and judiciary branches	Administrative system dominates, with weak legislature and judiciary, all influenced by the CCP
Federal and state governments are parallel systems of government	Hierarchical, pyramidal system of government

*China has five levels of government: central (中央), provincial (省级), prefecture (地级), county (县级), and township (乡镇级). The city has two types: prefecture city (地级市) and county city (县级市). The top leader of the former is at the bureau level (局级) and the top leader of the latter is at the division level (处级) in the rank of government officials (please see table 1-1 in *China Statistical Yearbook 2016* [National Bureau of Statistics of China 2016]).

selected leaders. Both countries permit the existence of multiple parties, but in the United States the Republican and Democratic parties dominate and in China the Chinese Communist Party (CCP) has overwhelming control. The United States has parallel structures of government at the federal, state, and local levels, but China's structures are strongly hierarchical. The separation of powers is one of the hallmarks of the US Constitution, but in China the administrative system dominates a weak judiciary and legislature.

The most important distinguishing difference between the policymaking structures of the United States and China is the role of political parties. In the United States, political parties serve important roles in elections at all levels, which means that for American political parties, their most important task is to help their candidates get elected so they can occupy various positions in the government. The organization and the control of members is very loose for US political parties. American citizens can join or exit parties very easily, or remain "unaffiliated" or independent from a party but still be allowed to vote. In short, American political parties do not directly control the government. It is unimaginable that the Republican or Democratic Party would literally draft policies that are then directly implemented by government officials.

The role of the CCP in China is totally different from its counterparts in the United States. The CCP is a Leninist party with strict discipline and rigorous organization. No one who joins the party will never be allowed to exit voluntarily from it. The CCP's influence on government affairs is much broader and deeper compared with political parties in the United States, and the Chinese government in many ways acts as a functional branch of the CCP. We will discuss the role of political parties in much more detail in the following sections, and in chapter 6, but we wanted to make this fundamental distinction clear from the start.

American Structures for Policymaking

In the United States, the policymaking system (outlined in table 3.2) is strongly affected by the separation of powers between Congress (the legislative branch), the president (the executive branch), and the court (the judicial branch) at both the federal and state levels. In general, the legislature is empowered by the Constitution to write and pass new laws, which then must be signed by the president (and most governors) for them to enter into force. If the president vetoes a bill, it returns to the legislature for reconsideration. These laws designate authority to government agencies overseen by the US president or a state governor to implement and enforce laws. Then, the relevant government agencies either directly implement provisions of the laws or develop rules and regulations to achieve the law's objectives. These government agencies also are responsible for enforcing the laws that are passed by the legislature. If any other level of government or affected stakeholder believes that a law is not consistent with the Constitution or that a new regulation is illegal under an existing law, they can challenge these laws in the courts. The state and federal courts will hear these challenges and try to resolve them at their level through rulings. The legal decisions at the state and federal level can be appealed to a higher-level court all the way up to the Supreme Court if the dispute concerns a federal law or an issue related to the US Constitution.

Below the level of the fifty individual states, there is considerable variation in the organization of the policy process, and these decisions were left to the states when the Constitution was originally written. Most states

Table 3.2

US levels of government and basic policymaking authorities

	Legislative	Executive	Judicial
Federal	US Congress	US president Federal government agencies	Supreme Court Federal Court
Authorities	*Develop and pass laws*	*Implement and enforce laws*	*Resolve legal disputes about the meaning of laws and whether they violate the Constitution*
State	Legislature	Governor State government agencies	State courts
Authorities	*Develop and pass laws*	*Implement and enforce laws*	*Resolve legal disputes about the meaning of laws and whether they violate the Constitution*
Local	Government organization varies below the state level; examples provided		
County	*Example:* legislature	*Example:* county administrator	*Example:* county court
Local (city or town)	*Examples:* town meeting, city council	*Examples:* mayor, selectman	*Examples:* municipal court

have county-level governments, which are effectively regional divisions within the states. Below the county level are individual cities and towns—municipalities—that also have the ability to create and implement local policies and rules through various governance arrangements permitted under state law. Local towns and cities usually have considerable power over educational policy, building codes, and local economic development.

The governance arrangements vary dramatically at the substate level and generally have some regional characteristics. In New England, for example, it is common for local towns to have direct responsibility for services, with a highly democratic system of elected town member representatives and a *first selectman* (similar to an elected mayor), who come together at town meetings to make administrative decisions about local schools, zoning regulations, libraries, waste disposal, and so forth. In the more rural Midwest, it is more common to rely on elected officials of counties rather than towns for the provision of local services, with counties providing the schools, courts, libraries, fire, and police or sheriff departments.

In climate change policy, all levels of US government are important. To date, the most impactful policies to reduce greenhouse gas emissions have been at the federal and state levels, but some county- and local-level policies have been much more progressive than state or federal policies. These local leaders often show state and federal governments what is possible, experimenting with new policies. Likewise, state-level policies often demonstrate to the federal government what is possible at the federal level. For adaptation policies, county- and local-level policies have proven to be at least as influential as federal policies.

At the federal level, the policy process begins when the US Congress passes a new law that is signed by the US president. The law is then codified in the United States Code. Typically, a new law will delegate authority for implementing the law to one or more federal government agencies in the executive branch. For climate change policy, most of the relevant laws delegate authority to the US Environmental Protection Agency (EPA), US Department of Energy (DOE), US Department of Transportation (DOT), US Department of the Interior (DOI), US Department of Agriculture (USDA), and various science agencies, including the National Oceanic and Atmospheric Administration (NOAA) under the US Department of Commerce (DOC)[2] and the National Aeronautics and Space Administration (NASA).

These agencies then are responsible for implementing the law. In some cases, the law will have specific provisions that can be directly implemented and enforced. In other cases, the law directs the federal agency to develop specific regulations or rules to meet the objectives of the law. The Clean Air Act, for example, specifically directed the EPA to implement an emissions-trading scheme for sulfur dioxide but more generally charged the agency with developing and updating national air quality ambient standards that are necessary to protect public health (US Code 2017a). The agency will then propose a regulation and receive public comment on its proposal. The proposal is listed in the Federal Register. Once the agency reviews the public comments on the proposed rule, it will then revise the rule and issue a final regulation. The final rule is also published in the Federal Register and is codified in the Code of Federal Regulations (EPA 2017a).

The agencies also have direct responsibility for enforcing the laws and regulations. The law will specify whether or not violations are a criminal or civil matter. For a criminal prosecution, the violator can be sent to prison. For a civil prosecution, the violator can pay a large fine or be required to correct the violation or to take additional corrective action (EPA 2017a). The federal agencies directly enforce federal regulations even at the local level. Thus, federal regulators work in parallel with state regulators to regulate the same factories and power plants within a state.

The federal system described in the preceding paragraphs is generally replicated at the state level. State laws have equal standing with federal laws, and thus American citizens and corporations are subject to two parallel systems of policy. The states do not need to obtain the approval of the federal government to promulgate new laws or regulations, but of course they must not contravene federal laws. Below the level of the state, however, it is impossible to generalize about the policy process in the United States due to the variation at the county and municipal levels (Rabe 2004).

In climate change policy, local governments are more likely to regulate their own emissions by establishing local targets for emission reduction, changing the practices of a municipal electric utility (if they have one), providing public transit and zoning laws, and through changing building codes to require greater levels of energy efficiency. New York City created its own energy code in 2009 to boost the energy efficiency of its buildings, for example, and estimates an annual savings from this and other new local building policies of 3.4 million metric tons of CO_2eq when all new policies in the One City: Built to Last plan are implemented, plus a cumulative financial savings of USD 8.5 billion over ten years (New York City Government 2014, 95–98). Local governments also are issuing new adaptation policies and incentives through their zoning policies and public-private partnerships (PPPs). New York City, for example, established the CoolRoofs program in 2010 as a PPP in order to encourage coating city rooftops with a white, reflective coating to reduce roof and building temperatures in the summer and increase reflectivity of sunlight all year round. As of August 2016, more than six million square feet had been coated (New York City Government 2016).

Chinese Structures and Formats for Policymaking

In this section, we introduce the structure of China's policymaking system, emphasizing the hierarchical, pyramidal structure of China's government.[3] This structure is depicted in figure 3.1, wherein the solid lines represent formal lines of authority and the dotted lines represent more informal or hidden relationships.

The main two differences we wish to emphasize in this section relate to the *levels* of government and the *role of the CCP*. Constitutionally, the National People's Congress (NPC; 全国人大) is the highest organ of state administration, as depicted in figure 3.1, and theoretically it supervises the work of the State Council (国务院), the State Central Military Commission (中央军事委员会), the Supreme People's Court (最高法院), and the Supreme People's Procuratorate (最高检察院). The heads of each these entities technically must be approved by the NPC. This constitutional arrangement belies the pervasive influence of the Chinese Communist Party in all matters, and the head of the NPC is actually ranked third in the party hierarchy, after the president and the premier. Indeed, the preamble of the State Constitution declares that the work of the Chinese people will take place "under the leadership of the Chinese Communist Party" (NPC 2004).

The central government structure depicted in figure 3.1 is generally replicated at each level of government in China. In the United States, federal, state, and local governments promulgate and implement policies in parallel, without need for approval from higher levels of government. In China, the system is vertically hierarchical, and lower levels of government must have the consent of higher levels of government for major new policies, even though lower levels can issue new policies and laws that complement central government policies.

According to Article 89 of China's Constitution, the State Council "exercises unified leadership over the work of local organs of state administration at various levels throughout the country, and formulates the detailed division of functions and powers between the Central Government and the organs of state administration of provinces, autonomous regions, and municipalities directly under the Central Government." In most cases, when the central government releases an important policy document, the provincial government will release a corresponding document, which reflects the central government's document and its own specific situation. In the climate change arena, for example, the central government released "The Work Plan for Controlling Greenhouse Gas Emissions during the 13th Five-Year Plan Period" ("十三五"控制温室气体排放工作方案; State Council 2016a) and then the Guangdong Province released its own document, "Guangdong Work Plan for Controlling Greenhouse Gas Emissions during the 13th Five-Year Plan Period" (广东省"十三五"控制温室气体排放工作实施方案; Guangdong 2017).

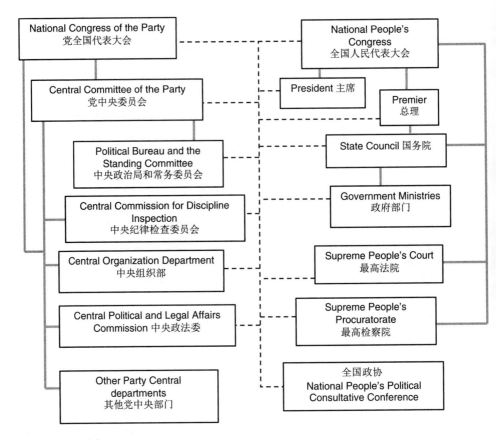

Figure 3.1
Chinese government organization at the central level. This figure depicts the organization of the Chinese government at the central level. Constitutionally, the National People's Congress (全国人大) is the highest organ of the policymaking structure, but according to the Constitution of the Chinese Communist Party (中国共产党章程), "the Party commands the overall situation and coordinates the efforts of all quarters, and the Party must play the role as the core of leadership among all other organizations at corresponding levels." In practice, therefore, the Communist Party influences every government institution. Each organization portrayed here is replicated at each level of government (provincial, city, county).

CPC (Communist Party of China: 中国共产党): The party plays a dominant role at each level.

National Congress of the Party (党全国代表大会): The highest leading body of the party.

CPC Central Committee of the Party (党委): Holds the highest power in the CPC at each level. At the central level, the committee has about two hundred members (中央委员) and 170 alternate members (中央候补委员; currently, it has 209 persons and 172 persons, respectively).

Figure 3.1 (continued)

Political Bureau and the Standing Committee: The top leaders of the Party Committee (at the central level, there are twenty-five persons in the Political Bureau [中央政治局] and seven persons in the Standing Committee at present [中央政治局常委]).

Central Commission for Discipline Inspection (纪委): Disciplines party members.

Central Political and Legal Affairs Commission (政法委): Leads and coordinates the organs of public security, the procuracy (检察院), and the court (法院).

Central Organization Department (组织部): Manages the high-rank officials. At the central level, it manages the vice minister and above.

People's Congress (人大): In theory, the people's congresses at all levels are constituted through democratic elections, and they are responsible to the people and subject to their oversight.

Government (政府): Led by the premier, and containing ministries such as the NDRC (发改委), MOF (财政部), and MOST (科技部).

Political Consultative Conference (PCC; 政协): In theory, the main functions of the PCC are political consultation, democratic supervision, and participation in the deliberation and administration of state affairs.

People's Court (法院): In theory, the court is independent.

Laws and regulations in China can be divided into two basic types. The first type is laws made and promulgated by different levels of people's congresses (各级人民代表大会). These include national laws (国家法律) enacted by the National People's Congress and the Standing Committee, as well as laws created by local decree (地方性法规), autonomous decree (自治条例), and special decree (单行条例).[4] Provinces, autonomous regions, municipalities directly under the central government and the Standing Committee, and major cities enact these decrees. The second type of law and regulations are promulgated by the administrative branch of the government, and these include administrative regulations (行政法规) that are enacted by the State Council, administrative rules (部门规章) that are issued by agencies under the State Council, and local rules (地方规章) promulgated by local governments. The law- and regulation-making system in China can thus be categorized as depicted in table 3.3, from top to bottom (NPC 2000:Legislation Law立法法).

Although work and consultation toward comprehensive national climate change legislation are still underway, climate-relevant decrees have been issued at the subnational level. For instance, the provinces of Qinghai and Shanxi have taken the lead to release Measures on Addressing Climate Change respectively in 2010 and 2011, and the Provisions of Carbon

Table 3.3

The law- and regulation-making system in China

Organization	Laws and regulations
National People's Congress (全国人大)	National laws (国家法律) such as criminal, civil, and state organic laws and other basic laws
Standing Committee of the National People's Congress (全国人大常务委员会)	National laws (other than those to be enacted by the National People's Congress within state legislative power)
State Council (国务院)	Administrative regulations (行政法规)
Ministries (国务院各部门)	Administrative rules (部门规章)
People's Congress of a province and the Standing Committee (省级人大及其常委会)	Local decrees (地方性法规)
People's Congress and its Standing Committee in a major city (较大市的人大及其常委会)	Local decrees (地方性法规)
People's Government (地方政府)	Local rule(地方政府规章)

National Law (国家法律): National law may be enacted in respect of matters relating to (i) state sovereignty; (ii) the establishment, organization, and authority of various people's congresses, people's governments, people's courts, and people's procuratorates; (iii) the autonomy system of ethnic regions, system of special administrative region, and system of autonomy at the grass-roots level; (iv) crimes and criminal sanctions, etc. (Legislation Law: Article 8).

Administrative Regulation (行政法规): Administrative regulations may provide for the following: (i) matters for which enactment of administrative regulations are required in order to implement a national law; (ii) matters subject to the administrative regulation of the State Council under Article 89 of the Constitution (Legislation Law: Article 56).

Administrative rules (部门规章): A matter on which an administrative rule is enacted shall be a matter which is within the scope of implementing national law, administrative regulations, and decisions or orders issued by the State Council (Legislation Law: Article 71).

Local Decree (地方性法规): A local decree may provide for (i) matters in order to implement a national law or administrative regulation in light of the actual situation of the jurisdiction; (ii) matters that are local in nature and require the enactment of a local decree (Legislation Law: Article 64).

Local Rule (地方政府规章): A local rule may provide for the following: (i) matters for which enactment of local rules is required in order to implement a national law, administrative regulation or local decree;(ii) matters that are within the regulatory scope of the local jurisdiction (Legislation Law: Article 73).

Emissions Management of the Shenzhen Special Economic Zone, passed by Shenzhen Municipal People's Congress in 2012, also forms the legal basis for the regional emissions-trading system (ETS) in Shenzhen (Shenzhen People's Government 2014). The Shenzhen ETS is a good example of an administrative regulation.

Policy documents, typically released by party committees and governments at all levels, dominate China's climate policy landscape. The assorted policy documents across different sectors and at different levels form the backbone of the nation's climate change policies. These policy documents can be classified into four general levels based on their issuing agency, as depicted in table 3.4.

First and foremost, major documents (documents of great importance), such as "The Decision of the Central Committee of the Communist Party on Some Major Issues Concerning Comprehensively Deepening Reforms" (Xinhua 2013) adopted at the Third Plenary Session of the Eighteenth CPC Central Committee and "Proposal for Formulating the 13th Five-Year Plan for National Economic and Social Development (2016–2020)," are promulgated by the Central Committee of the Chinese Communist Party and the National People's Congress. These documents are overarching guiding

Table 3.4
The document and policy file system in the climate domain in China

Level	Documents/ policy files	Organization/ official	Examples
1	Documents from the party	National Congress and the Central Committee of the party	Party Congress report, party decisions on reform or five-year plan
2	Documents from the government or top leaders' speeches	State Council or the top leaders	Five-year plan, remarks by the president or premier at the UN Climate Summit
3	Policy files	Ministries and commissions directly under the State Council	National Plan to Address Climate Change (2014–2020), China's National Climate Change Programme
4	Policy files	Subnational provincial, county, city, or local governments	Provincial Work Plan for Controlling Greenhouse Gas Emissions during the 13th Five-Year Plan Period

documents to direct the future development and reform of China. Contained in these documents are common understandings and aspirations of the entire party, and they are intended to encompass the wishes of the people of the entire country. Touching upon every aspect of social development, including climate change, these documents identify and clarify future directions for policy and set the tone for China's action on climate change.

The second level of policies is embodied in documents issued by the State Council or expressed in speeches by the state leaders. Some examples of this level of policy include the thirteenth five-year plan and remarks by the Chinese president or premier at the UN Climate Summit. They reflect the country's attitude and overall policy approach toward climate change. Of particular significance, they help to convey to the international community China's stance on climate change, as well as its general work plan.

The third-level policies are issued by ministries and commissions directly under the State Council, particularly China's primary government institution responsible for climate change governance, the National Development and Reform Commission (NDRC). These include the National Plan to Address Climate Change (2014–2020) (国家应对气候变化规划(2014–2020年) and the 2007 China's National Climate Change Programme (中国应对气候变化国家方案; Xinhua 2015e). These policies can be further classified into those issued by ministries and commissions (like the Ministry of Transport or Ministry of Housing and Urban-Rural Development) or their subsidiary departments (like the Climate Change Department of NDRC), with the former covering a broad area of topics and the latter focused on more specific issues. These policies usually target specific sectors overseen by the policy-issuing agency, though many of them may be indirectly related to climate change, such as in the sectors of transportation and housing construction. Elements of their titles, such as Measures, Opinions, or Notices, also indicate different contents and the nature (such as whether they are compulsory or not) of these policy documents.

The fourth-level policies are those issued by subnational provincial, county, city, or local governments, as well as other local authorities particular to specific sectors to supplement central policies. Lower levels of government are obligated to implement and enforce the policies of higher levels of government. A prefecture-level city, for example, would need to implement the policies of its province and the central government in addition to any of its own policies.

In summary, China's climate change policies are composed of and driven by policy documents issued at different levels. These policies are distinguishable mainly by the level of their issuing agency. Policies promulgated by different levels of government agencies are different in terms of objective and positioning. Although principles and overall directions are usually the focus of policy documents issued by the central government, those released by local governments focus more on the implementation details.

Another attribute of China's climate policy framework is its flexibility. Given the prevalence of policy documents in the framework, China's climate change policies are not standardized, nor are they highly prioritized at the subnational level. Their implementation and impact often varies around China, depending on the timing, location, and people in charge. Their flexibility also originates from the fact that most of the time the policy documents only mention the principles, leaving much more room for interest-based bargaining in the implementation process. It is reasonable to some extent to let local governments consider their own situations and actual needs given the great disparities among different regions in China. On the other hand, specific quantitative targets set forth in central government guidelines or plans must be met somehow.

The aforementioned characteristics are interwoven with the administratively driven policy processes, centralization of authority, and central-local relations that feature in China's political system. Governments at all levels of the administrative system are the main bodies to formulate and implement climate change policies, whereas other organizations (such as the National People's Congress) have a relatively weak role to play in the policymaking process. Although major laws are indeed passed by the NPC, these laws are the foundation for policy documents that are developed and promulgated by the administrative state. These laws only become effectual when they are supplemented by administrative rules (部门规章) and polices such as the Renewable Energy Law and Energy Conservation Laws released by the NPC. These laws are more principles, frameworks, not the detailed measures that can be implemented directly in reality. In the process of the making of Renewable Energy Law, the NPC (全国人大) asked the NDRC (国家发改委) and MOF (财政部) to also release some related operable rules and polices (Yilin 2016). To date, the Supreme Court has played an unimportant role, but, after all, there is no national law on climate change in China, so there can be no court cases. In other related areas, such as environmental protection or transportation, the judicial branch has heard court cases, but it is estimated that a very small proportion of environmental disputes actually are resolved in the court system (Economy 2014). So, to a great extent, China's policy framework on climate change has its origin in the existing central-local relations. The central government, as the center of authority and at the top of the hierarchy, initiates and formulates the majority of climate policies (particularly the significant ones), leaving implementation to local governments. As a result, the ultimate impact of these policies will depend both on the design, supervision, and examination of these policies to be undertaken by the central government, and on effective support to be provided by the local government in the process of policy implementation.

Comparing Major Actors

In terms of major actors, as depicted in table 3.5, at the federal or central government level the US Congress and Supreme Court play much more important roles than their counterparts do in China. Conversely, the CCP has much more power than do the Republican and Democratic parties in the United States.

Another major difference is the existence and influence of state-owned enterprises (SOEs) in China (see box 3.1). If a leader of an SOE has a

Table 3.5
Major actors

United States	China
President, Congress, Supreme Court	Party (党), National People's Congress (人大), People's Political Consultative Conference (政协), State Council (国务院)
Federal, state, and local departments and agencies	Central, provincial, and local government ministries and bureaus
Democratic and Republican parties	Chinese Communist Party
n/a	State-owned enterprises (SOEs)
Private sector	Private sector
Nongovernmental organizations (NGOs)	Civil society organizations
Experts	Experts

Box 3.1

The state-owned enterprise (SOE) system in China

SOEs are organized into different levels, similar to how the government is structured in China. At the central level, SOEs are owned by the central government and are called central SOEs (央企). The top leaders of central SOEs are managed by the Central Organization Department of the party (中组部) and SASAC of the State Council (国务院国资委), if they rank vice minister (副部级) or above. Lower-rank employees in the Central SOE are managed by the human resource departments in the company, and some, like professional managers, are hired in the market and may be dismissed if market conditions change.

Similarly, provincial SOEs (省属国企) are owned by provincial governments (省政府), and the top leaders of the Provincial SOEs are managed by the provincial Organization Department of the party (省委组织部) and the SASAC of the provincial government (省国资委), if they rank vice bureau (副局级) or above.

The prefectural SOEs (市属国企) are owned by prefectural governments, and the top leaders of the prefectural SOEs are managed by the Prefecture Organization Department of the party (市委组织部) and the SASAC of the prefecture government (市国资委), if they rank vice division (副处级) or above.

Level (级别)	SOE	Rank of some top leaders	Organization department (party) 组织部 (党)	SASAC (国资委)
Central (中央)	Central SOE (央企)	Vice minster (副部级) or above	Central Organization Deptartment (中组部)	SASAC of State Council (国务院国资委)
Provincial (省级)	Provincial SOE (省企)	Vice bureau (副局级) or above	Provincial Organization Department (省委组织部)	SASAC of provincial government (省国资委)
Prefecture (地级)	Prefectural SOE (市企)	Vice division (副处级) or above	Prefectural Organization Department (市委组织部)	SASAC of prefectural government (市国资委)

vice-minister-level ranking (副部级) or above, he or she is appointed by the Central Organization Department of the CPC (中组部) according to his position in the party. The State-owned Assets Supervision and Administration Commission of the State Council (SASAC) (国资委) will simultaneously appoint the top leaders of an SOE on its own because it is responsible for the leader's position in the company. To be clear, the real power is in the CCP Organization Department based on the principle that the party is in charge of cadres (党管干部原则), so the SASAC appointment is just a procedure. In fact, there is a party committee (党委) or party group (党组) within each SOE, and usually the secretary of the party committee or group (党委书记 or党组书记) is the chairman of the board (董事长), such as is the case with PetroChina (中石油), for which Wang Yilin (王宜林) is simultaneously the chairman (董事长) and the secretary of the party group (党组书记) in the company. Wang Yilin's appointment was announced by the vice minister of the Central Organization Department of the CCP (People's Daily 2016). Some top-leaders in very important central-level government-owned enterprises (央企) occupy high positions in the party. The leaders of the State Grid Corporation of China (SGCC; 国家电网), Sinopec (中石化), and China National Nuclear Corporation (CNNC; 中核集团) are alternate members of the central committee of the CPC (中央候补委员), which means they are ranked within the top three hundred highest positions in China.

As such, high-ranking SOE leaders are in a position to be involved in the policy process in a way that is unparalleled in the United States. Usually, the SOEs are not directly involved in the policymaking process, but they will give advice and feedback to the government. In the past, in some special sectors such as petroleum and electricity, SOEs like PetroChina (中石油), Sinopec (中石化), or SGCC (国家电网) were more powerful and capable than the government agencies responsible for them, so they were deeply involved in the policymaking process.

It is important to point out that the ownership of the SOE greatly matters because only SOEs owned by the central government can influence central government policy. SOEs owned by local governments will have a strong influence on local government policy, but it is very difficult for them to influence higher levels of government. At the subnational level, the US states and Chinese provinces both develop policy, but the federal government in the United States enforces its own policies, whereas the provinces and local governments in China must enforce central government policies as well as their own.

As discussed in more detail ahead, nongovernmental actors have more influence on the policy process in the United States compared with China, where civil society has a muted and indirect role. In the United States, NGOs including the Natural Resources Defense Council, Environmental Defense Fund, Greenpeace, Sierra Club, World Wildlife Fund, Union of Concerned Scientists, and many others employ staff scientists and lawyers as in-house experts. Many of these organizations have lobbying arms as well. In China, civil society organizations are largely limited by government regulations to educating the public (Economy 2014; Ma and Jia 2015).

In both countries, the private sector tries to affect the policy process, but it has more direct means to do so in the United States than in China (except for state-owned enterprises). Many of the private firms in the United States form lobbying groups such as the American Petroleum Council or the National Association of Manufacturers, which directly lobby on their behalf. In China, private entrepreneurs can impact policy process through informal means or by becoming representatives of the People's Congress (人大代表) or of the Political Consultative Conference (政协委员) at different levels of government. As representatives, they can submit policy proposals and advice using their status as representatives. Some of their advice and proposals have a chance of becoming new policies or at least of impacting policymaking processes.

Finally, outside experts can influence governments in both countries, but it is more customary and less formal in China than in the United States, in part because of a US law, the Federal Advisory Committee Act (FACA), which was enacted in 1972. This law was intended to ensure that advice received by experts was objective and accessible to the public. According to the law, "the records, reports, transcripts, minutes, appendixes, working papers, drafts, studies, agenda, or other documents which were made available to or prepared for or by each advisory committee shall be available for public inspection and copying at a single location" (GSA 2000). This provision, among others, makes it cumbersome and difficult for groups of experts to be convened and to convey their advice to decision-makers. Any advisory group, with limited exceptions, that is established or utilized by a federal agency and includes at least one member who is not a federal employee must comply with FACA (GSA 2017). As a result of these bureaucratic procedures, many US federal agencies will forgo the opportunity to solicit advice on an ad hoc basis through expert groups because it cannot be done in a timely way or it is too cumbersome and bureaucratic to do so. Only those advisory bodies that already exist are likely to be turned to when and if agencies need expert input.

In China, groups of expert advisors are routinely convened to provide input for policy decisions within particular ministries or departments or as a cross-cutting matter. Scientific expert groups advise the Ministry of Science and Technology on particular aspects of technological innovation policy, for example, and are usually organized by type of expertise. On the overall topic of climate change, a National Panel on Climate Change (国家气候变化专家委员会) was convened to directly advise the National Leading Group to Address Climate Change (国家应对气候变化领导小组), which is comprised of relevant ministers within the State Council (Hart et al. 2015; Lewis 2008), but also to submit suggestions to other government agencies, including the NDRC (国家发改委) and State Meteorological Administration (国家气象局) (CMA 2011). Other expert organizations, including the Development Research Center of the State Council (国务院发展研究中心), the Chinese Academy of Sciences (中国科学院), top universities (especially Tsinghua University in the case of climate change), the Energy Research Institute (宏观经济研究院能源研究所), and the National Center for Climate Change Strategy Research and International Cooperation Center (NCSC: 国家应对气候变化战略研究和国际合作中心) under the NDRC, are among the top think tanks that advise the Chinese government on climate policy.

Comparing Government Processes and Approaches

As hard as it is to synthesize the different approaches to policymaking in the United States and China, we endeavor to do so in this section. Table 3.6 provides a high-level summary of the comparisons that are discussed in more detail in the text. We endeavor to make direct parallels in the table, but in some cases there are no direct parallels given the many disparities between the United States and China. Bargaining and coalition formation are essential to the policy process in both countries, for example, but the policymaking and enforcement process is completely unalike at different levels of government.

We first analyze US approaches to policy, highlighting key factors that are prominent features of the American policymaking system, before doing the same for China. At the end of this chapter, we highlight some of the advantages and disadvantages of each system.

American Approaches to Policy

The American policy process shifted dramatically, becoming more polarized and gridlocked due to party politics beginning during the Clinton administration of the 1990s. After the election of Democratic President Bill Clinton, the Republican Party (GOP), led by Speaker of the House Newt Gingrich, launched the "Republican Revolution" in 1994, which led to a net gain of fifty-four seats in the House of Representatives and eight seats in the Senate for the GOP, and resulted in the Republicans gaining control of both the House and Senate for the first time in forty years. Although President Clinton retained his constitutional veto power, it became much harder for him to advance and pass legislation. At the same time, the Republicans could not convince him to sign bills into law. Eventually the impasse grew so severe that the government literally shut down. Today, the GOP generally advocates for a smaller federal government, limiting immigration, tax cuts, eliminating waste in spending, and reductions in most of the social programs advanced by Democrats, including social security and welfare programs.

Focusing on environmental policy, Gingrich's close partner House Whip Tom Delay was a fervent antienvironmentalist who introduced legislation to revoke the ban on ozone-depleting and greenhouse gas chemicals in 1995. Delay repeatedly called for the elimination of the EPA, a refrain taken up by President Trump's administrator of the EPA, Scott Pruitt. With Delay began a new era of political polarization on environmental policy that continues to this day. Although environmentalism used to be mostly a bipartisan issue, the 1994 Republican Revolution marked the beginning of a split, with the Republicans being generally much more skeptical of the need to support climate policy than Democrats or independents. In general, the Republicans believe the federal government has overreached in regulations of all kinds. Perhaps by coincidence, Republicans also receive the majority of the fossil fuel industry's campaign contributions (discussed in detail in chapter 6).

Table 3.6
Comparing processes of the U.S. and Chinese policymaking systems

United States	China
Senate provides advice and consent for senior government appointments.	The CCP controls the selection of government officials; formal appointment of highest-level official is by the NPC.
Senior government officials come and go through a "revolving" door.	Government officials are promoted by the CCP.
Senate ratifies international treaties.	The NPC ratifies international treaties as a formality.
Consensus is not required within the executive branch in decision-making because authority rests with the president. Congress decides by majority or supermajority vote.	Consensus-building desired at all levels; decisions made by upper levels; difficult but not impossible for president (also the general secretary of the CCP) to make a sole decision.*
Bargaining and coalition-formation among stakeholders.	Bargaining and coalition formation among stakeholders.
Supreme Court provides a final decision about federal laws and constitutional matters, and can overrule both Congress and the president.	The court system is limited in power in China, but judicial activity is growing.
Public provides direct input to policy process.	Public can usually, but not always, express opinions.
Bureaucratic politics are intense.	Bureaucratic politics are intense.
Direct lobbying by interest groups.	Interest groups engage in official process and also indirectly influence through *guanxi*.
Individual and corporate campaign contributions are legal.	No official channel for financial influence, but corruption exists.
Informal elite networks of people can have great influence.	Informal elite networks of people can have great influence.
Members of Congress are directly elected. The president is elected through the Electoral College.	The Congress and Party Congress are constituted by elections,** but in reality, according to the principle of democratic centralism (民主集中制), the allocation of political positions mainly occurs through top-down methods.***
Leadership of Congress elected by Congress members.	NPC leadership formally elected by NPC members with party selection.

Table 3.6 (continued)

United States	China
Direct enforcement of federal policies, including use of courts.	Enforcement of central government policies by local level officials.
Bottom-up policy experimentation in the states.	Central government–initiated policy experimentation in which local governments participate on a compulsory or voluntary basis.

*The most senior of the top leaders in China simultaneously occupies three positions: president of the state (国家主席), the general secretary of the Central Committee of the CCP (中国共产党中央委员会总书记), and the chairman of the Central Military Commission (中央军事委员会主席). The president of the state (国家主席) promulgates statutes, appoints or removes the premier, vice premiers, state councilors, and ministers in charge of ministries or commissions in pursuance of the decisions of the National People's Congress and its Standing Committee (China Constitution: Article 80). The general secretary of the Central Committee of the CCP (中共中央总书记) is number one in the party. The chairman of the Central Military Commission (中央军委主席) is chief commander of the military. Xi Jinping currently occupies these three positions. The general secretary of the party and chairman of the Central Military Commission are more important, and the president of the state is more symbolic. In history, these three positions have been occupied by different people at same time, such as in the early 1980s when Hu Yaobang (胡耀邦) was the general secretary of the party, Deng Xiaoping (邓小平) was the chairman of the Central Military Commission, and Li Xiannian (李先念) was the president of the state. In the early 2000s, Hu Jintao (胡锦涛) was the general secretary of the party, and Jiang Zemin (江泽民) was the chairman of the Central Military Commission. The allocation of the positions reflects the political distribution of power.

**In Article 3 of China's Constitution, "the National People's Congress and the local people's congresses at various levels are constituted through democratic elections." And in Article 10 of the CPC Constitution, the "Party's leading bodies at all levels are elected."

***From a Xinhua report about how the new leaders of the nineteenth party Congress were selected (党的新一届中央领导机构产生纪实), it is clear that "selection by [a] top-down method" occurred rather than "election by [a] bottom-up method" (Xinhua 2017c).

The emergence of the Green Party is another important factor for climate change policy. Although the US national Green Party is itself relatively tiny compared to the Republican and Democratic parties, it may have swayed the 2000 presidential election after its formation in the mid-1990s. In 2000, Ralph Nader ran for president on the Green Party ticket against Vice President Al Gore, who was best known at the time for his advocacy on ozone depletion and climate change. Nader ended up winning only 3 percent of the national vote, but it may have been enough to affect the electoral college outcome. Alternatively, it is possible that fear of losing votes to Ralph Nader caused Gore to run further to the left politically, which meant he lost votes to George W. Bush (Erikson 2001). Few believe that the Green Party was a decisive factor in Hillary Clinton's loss to Donald Trump, but many younger voters did not support her at the same level they did Barack Obama, and one-tenth voted for third-party candidates, including Jill Stein of the Green Party (Mosendz 2016).

The US Electoral College is puzzling to many people, inside and outside the United States. How is it possible that a candidate that wins the popular vote, as Al Gore did in 2000 and Hillary Clinton did in 2016, does not become president? The Electoral College was established in the Constitution as a compromise between election by popular vote of citizens and a vote by the US Congress. Each state is allotted a number of electoral votes, and political parties in each state select the electors. The process varies state by state. Electors generally are not required by federal law to vote consistent with the results of the popular vote in each state, but some states require them to do so. Most states have a winner-take-all system that awards all electoral votes to the winning presidential candidate, but Maine and Nebraska do not (Archives 2017). Getting rid of the Electoral College system would require an amendment to the US Constitution and would be very difficult to accomplish given the political dynamics in Congress.

Central to the US policy process are the many checks and balances built into the Constitution to prevent any one individual, US state, or branch of the federal government from becoming too powerful. The Constitution grants certain rights and responsibilities to the federal government and the states. The federal government is responsible for the defense of the nation, and it collects the majority of the tax revenue to redistribute to the states through Congressional appropriations. The federal government itself has checks and balances built into it with three separate branches of government—the executive branch led by the president, the legislative branch embodied as the Congress, and the judicial branch led by the US Supreme Court. Even within each branch there are checks and balances— for example, a bicameral legislature, in which the Senate has two representatives from each state and the number of elected representatives from each state in the House is proportional to the population of each state. The Supreme Court has seven individual members who are nominated by the president when there is a vacancy, confirmed by the Senate, and serve for their entire lives unless they choose to step down.

Because of the federalist system of government, the federal government can enforce national laws and regulations directly at the state and local levels. The EPA, for example, enforces its policies at the state and local level. The EPA divides the United States into ten regions, and it has local offices in each region that are responsible for the implementation and enforcement of EPA policies and programs.

In the United States, *public engagement* is sacrosanct to the policy process. The government is constantly meeting with "stakeholders," holding hearings to solicit expert or stakeholder comment on legislation or important issues, and inviting public comment on new regulations and laws. Indeed, the legal procedure for most regulatory processes requires public engagement. For the Obama administration's Clean Power Plan rule, the EPA received 4.3 million comments from the public during the six-month comment period and was required to respond in writing to each one before issuing the final rule (EPA Archive 2015b). These comments often do alter policy. In the Clean Power Plan, for example, EPA revised the timeframes for compliance to provide "glide paths" rather than "cliffs," to make the policy more flexible. Specifically, EPA decided that performance rates could be phased in over a 2022 to 2029 interim period, which leads to a glide path of reductions achievable on average over the eight-year interim period (EPA Archive 2015a, 2015c).

Within the executive branch, bureaucratic politics reign supreme. Of all the federal government agencies, the Department of Veteran Affairs has the largest number of employees—337,683 at last count—and the federal government as a whole has 2.8 million staff members (BLS 2017). The president's cabinet is comprised of the vice president and fifteen agency leaders, all appointed by the president and confirmed by the US Senate. In addition, seven other individuals have cabinet rank, including the administrator of the EPA, the US trade representative, and the ambassador to the United Nations. These leaders debate policies in cabinet meetings and compete for influence with the president. The bureaucrats within the agencies themselves compete with each other for budget authority, over policy priorities, and for attention from the White House.

Of all the theories proposed about the American policy process, the ones that seem to apply best to the US climate policy process are those advanced originally by political scientists Graham Allison and Morton Halperin (1972) and subsequently in 1984 by John Kingdon (2010). Allison and Halperin proposed their bureaucratic politics model of government decision-making in 1972. They asserted that "what a government does in any particular instance can be understood largely as a result of bargaining among players positioned hierarchically in the government" (Allison and Halperin 1972, 43). The organizational processes and shared values within the bureaucracies create constraints on the policy decisions.

In the United States, most senior officials are nominated by the president to head departments or agencies as political appointees and must be confirmed by the Senate. Political appointees thus have very short time horizons about policy decisions, whereas career civil servants often have a much longer time horizon regarding the interests of their organizations and are more likely to make decisions grounded in their bureaucratic interests and perspectives. Career civil servants go through regular performance evaluations, but this is less common for political appointees. In fact, political appointees serve at the pleasure of the president, and if he or she is unhappy with their performance, they will be asked to resign. These political appointees usually supported the campaign either financially or as policy advisors (or both) and thus are trusted to be loyal to the president's interests and priorities. Senior appointments, such as cabinet secretaries, are often long-standing friends or colleagues of the president or major donors to his campaign. Their loyalty is prized because the president seeks to ensure that his or her wishes are carried out in every agency of the government. Allison and Halperin note that more junior officials are often motivated by the desire to participate in a decision just to get into the "game"; as a result, they are likely to take stands that gain them access. Personal ambitions can also influence positions that officials take—for example, a desire to be promoted within the organization or to a senior position elsewhere in the government. They write, "Reputation depends on one's track record, thus players consider the probability of success as part of their stake" (Allison and Halperin 1972, 50).

Individual interests are combined with bureaucratic ones in the various policy and decision "games" that are played by government officials: "Organizations compete for roles and missions" and "rarely take stands that require elaborate coordination with other organizations" (Allison and Halperin 1972, 49–50) because they would rather have full authority and control over a particular mission. Sometimes an issue arises because a player sees something that he wants to change, and then moves to take action, but more often, the game is begun in response to a deadline or an event that creates an incentive to act. As President Obama's first chief of staff famously said about the economic recession, "Never let a serious crisis go to waste" (Wall Street Journal 2008). When he becomes aware that a game has begun, each player must determine his stand and then decide whether to play (if he has a choice) and, if so, how hard (Allison and Halperin 1972, 50).

Bargaining among individuals from the various government agencies is inherent to the policy process. For those in the process, bargaining advantages "stem from control of implementation, control over information that enables one to define the problem and identify the available options, persuasiveness with other players (including players outside the bureaucracy) and the ability to affect other players' objectives in other games, including domestic political games" (Allison and Halperin 1972, 50). Control of components of the policy process can yield considerable advantage. An individual can wield enormous power if he can determine who is invited into a particular process, who can attend which meetings, who will draft decision memos, and who will get credit for the policy decision and outcome. A study of policy networks in the British government led to the recognition that such networks can strongly affect policy outcomes.

Marginal and incremental progress can be punctuated by abrupt policy changes (Baumgartner, Jones, and Mortensen 2014). Advocacy coalitions can be one cause of major policy changes, along with external or internal shocks and policy-oriented learning. *Advocacy coalitions* are defined as "actors sharing policy core beliefs who coordinate their actions in a non-trivial manner to influence" a policy (Jenkins-Smith et al. 2014). The broader the coalition, the easier it is to argue that it represents the views of a wide swath of the American public. The American lore about unusual winning coalitions always harkens to the "Baptist and bootlegger" coalitions that advocated against sales of alcohol on Sundays. Churchgoing Baptists made the moral argument for abstinence at least one day a week, while bootleggers—those illegally selling liquor—quietly convinced politicians to keep laws in place (DeSombre 2000).

Beyond unusual coalitions, organized interest groups can wield enormous power in US policymaking. Interest groups can exploit the American commitment to public engagement through many channels: They can contribute funds to the campaigns of elected officials to try to influence their policies once elected. They can organize the contributions of public comments on regulatory processes or letter-writing campaigns to members of Congress regarding particular bills or pending legislation. They can request meetings with government officials to explain their concerns about or support for particular policies. And, of course, they can launch public education campaigns around particular issues to influence public opinion. In a study of 1,779 different policy issues, Martin Gilens and Benjamin Page (2014) examined the influence of different types of interest groups on US policy outcomes and determined that "economic elites and organized groups representing business interests have substantial independent impacts on U.S. government policy, while average citizens and mass-based interest groups have little or no independent influence" (Gilens and Page 2014, 565).

One major reason that elites and business groups can be so influential is because of the US laws regarding expenditures in support of political campaigns—otherwise known as campaign finance. Although the actual amounts that individuals, corporations (through political action committees), or party committees can directly contribute to political campaigns are limited, there are other ways that moneyed interests can affect elections. The 2002 Bipartisan Campaign Reform Act attempted to limit the amount of money that any organization could contribute to candidates running for office, but in an infamous court case known as *Citizens United v. the FEC*, the Supreme Court ruled in 2010 five to four that the Constitution's guarantee of freedom of speech prohibited the restriction of expenditures expressing "political speech" by corporations, unions, or other associations (Supreme Court of the United States 2010). As a result, corporations and other interest groups are now allowed to spend unlimited amounts of money on advertising and other media campaigns to influence elections. As an article in the *New York Times* noted, "The Supreme Court ... handed lobbyists a new weapon. A lobbyist can now tell any elected official: if you vote wrong, my company, labor union or interest group will spend unlimited sums explicitly advertising against your re-election" (Kirkpatrick 2010). Democrats have vowed to pass a constitutional amendment to supersede the Supreme Court decision.

Policy decisions typically reflect considerable compromise. "Compromise results from a need to gain adherence, a need to avoid harming strongly felt interests (including organizational interests), and the need to hedge against the dire predictions of other participants" (Allison and Halperin 1972, 53). Compromise must be achieved among lobbyists, interest groups, bureaucratic actors, and even branches of government for the passage of major new legislation.

After Allison and Halperin published their seminal work on bureaucratic politics, John Kingdon conducted multiple case studies of US federal policy processes and concluded that problems, policies, and politics "stream" through the policy system and that "policy windows" emerge at certain points of time that are seized by policy entrepreneurs, who then bring the three streams together to advance certain policies. *Policy entrepreneurs* can be individuals, groups of people, or even organizations. The Multiple Streams Approach (MSA), as it is known, does not assume that actors are purely rational or that policymaking is driven by persuasion or the social construction of identity and meaning, but rather that it is a process of deliberation between competing groups in which each advances reasonable arguments in hopes of persuading the decision-maker (Zahariadis 2014).

Out of the universe of possible problems that policymakers might choose to address, how is it that one issue rises to the top of the list of priorities? In the "garbage can" model of organizational choice, it is recognized that in any given organization—or government—people come and go and they are distracted by competing ideas and multiple demands on their time. As a result, they do not share goals, and thus problems are resolved through bargaining. People in the government dump various problems and possible solutions into a virtual "garbage can" over time. In this theory, opportunities for decisions arise through resolution, oversight, or flight. Problems solved by resolution come about through explicitly working to resolve the problems over time. Policy decisions occur opportunistically when problems are attached to other choices that come along if there is sufficient energy to make the new choice quickly. Choices that have been unsuccessfully associated with certain problems can become more attractive to a different problem that comes along, enabling a decision when solved through flight. In short, "Choices are made only when the shifting combinations of problems, solutions, and decision makers happen to make action possible" (Cohen, March, and Olsen 1972, 16).

In sum, windows of opportunity emerge for new policies to be created and implemented. These windows can only be created if a problem is politically accepted to be caused by human behavior (Stone 1989). It may be that the politics around a particular issue shifts or that one issue is traded off for another or that the public attitudes change, resulting in a demand for new types of policies. In negotiation theory, this phenomenon is called *ripeness*—the point in time when a negotiation is ripe like a juicy piece of fruit for a negotiated resolution (Zartman 2000). Similarly, policy breakthroughs usually occur when the timing is ripe. A window of opportunity seemed to emerge in the United States on climate change during President Obama's first term in office, but not quite enough votes could be mustered in the Senate to pass comprehensive climate legislation in the US Congress. Perhaps the issue was not yet ripe enough, perhaps the approach— emissions trading—was not the best one, or perhaps insufficient political capital on the part of President Obama was invested in persuading senators to vote for the bill.

Chinese Approaches to Policy

Kenneth Lieberthal and Michel Oksenberg's groundbreaking *Policy Making in China: Leaders, Structures, and Processes* (1988) provided the first external, comprehensive, and penetrating analysis of the policymaking process in post–Cultural Revolution China, and many of the observations from their book (and Lieberthal's later works) remain astute and relevant today. Many other Western scholars have continued to observe and analyze China's policy process since then (Barnett 1985; Fewsmith 2013; Gallagher 2006; Hart et al. 2015; Kong 2009; Lampton 1987; Lieberthal 2004; Lieberthal, Li, and Yu 2014; Mertha 2009; Saich 2011), and Chinese scholarship is emerging on this topic as well (Yu 2016; Zhu 2008).

The role of the Chinese Communist Party (CCP) in the policy process cannot be understated. It is the most important actor in China's policy process because it leads the government and is responsible for maintaining national order and security. Political considerations thus dominate policy decisions (Yang 2014; Yu 2014). The CCP dominates the state apparatus through its control of personnel appointments, its setting of policy guidelines (Lieberthal and Oksenberg 1988, 5), and the ultimate decision-making power of the Politburo Standing Committee (Zhou 2014). The CCP's political legitimacy is strengthened by being a "development-oriented state" (Yang 2014, 268).

Notwithstanding the fact that China is considered to be a one-party state, other political parties do exist, and eight parties play a role under the Chinese People's Political Consultative Conference (CPPCC).[5] The CCP wants to demonstrate that it tolerates the other parties (in part so it can use their power), so the CCP allows a sizable percentage of non-CCP members in the government. The CCP thus encourages the existence of certain political parties and prominently promotes some of their leaders. Minister Wan Gang (万钢) of the Ministry of Science and Technology (科技部), for example, is the chairman of the China Party for the Public Interest (中国致公党). The former minister of health (卫生部),[6] Chen Zhu (陈竺), was chairman of the Chinese Peasants and Workers Democratic Party.

The CCP is not like a Western, secular, political party, but rather should be thought of as a traditional hierarchical bureaucratic system. The CCP plays multiple roles, including those of public administrator, promoter of ideologies, and unifier of the nation. It can be considered an organizational "emperor" that rules in a way that is consistent with ancient Confucian culture but is also based on Marxist and Leninist beliefs and principles. Although the Chinese government is authoritarian, it considers itself to be moral and Confucian. Thus, in a patriarchal way, the government is responsible for the well-being of the entire nation and therefore is deserving of the respect and deference that would be provided to elder leaders of a family. Government officials can criticize ideas and oppose certain policies from higher levels of government in private, but they will risk being marginalized or losing promotions if they do so in public in the course of the policy development process (Zheng 2010).

The CCP diffuses throughout the entire Chinese government structure as can be seen in figure 3.1. Individual officials can and often do serve in multiple roles. Xi Jinping (习近平) is the general secretary of the party, the president of the state, and the commander in chief (中共中央总书记, 国家主席, 中央军委主席). Premier Li Keqiang (李克强), for example, is the premier, a member of the Politburo Standing Committee (政治局常委), and also the head of the State Council. In China, the State Council is comprised of various ministers of the different government agencies, but it is ultimately outranked by the Standing Committee of the Political Bureau, known as the "Politburo Standing Committee" (中央政治局常务委员会). According to the party hierarchy, the leader of the State Council, Premier Li Keqiang, is only one of the members of the Politburo Standing Committee and thus ranks below General Secretary Xi Jinping. The Politburo Standing Committee is a subset of seven to nine members from the bigger twenty-four-member Political Bureau (or "Politburo" 中央政治局). The Politburo and its Standing Committee are the ultimate decision-making body in China. The State Council coordinates the government bureaucracy. It is led by the premier and is comprised of the vice premiers and the leaders of twenty-nine ministries and commissions. The State Council is the executive body of the highest organ of state power and the highest organ of state administration according to Article 85 of the Chinese constitution. The local people's governments at various levels throughout the country are administrative organs under the unified leadership of the State Council according to Article 110 (CECC n.d.). See box 3.2 for more information about the relationship between central and local governments in China.

The administrative branch of the government, led by the premier and dominated by the CCP, is relatively more powerful than the NPC (全国人大) or Supreme Court (最高法院), and, in the case of climate policy, most important decisions to date have been made by government ministries, the State Council (国务院), and the Politburo (政治局) and its Standing Committee (常务委员会). The leader of the NPC (全国人大), Zhang Dejiang, is also a member of the Standing Committee (政治局常委). The National People's Congress has three thousand members and meets once per year. Although its influence has grown, it is subservient to the leadership of the Communist Party. China also has a Supreme Court called the Supreme People's Court (最高人民法院), but it has limited power and authority. Courts are not empowered, for example, to interpret administrative rules and regulations (SPC 2017b). Lower-level courts handle civil and criminal cases (CECC n.d.).

Next, we turn to the characteristics of the policy process in China. China's most important policy decisions are made through consensus building among the elite, in part because individual leaders have become progressively weaker (Lampton 2014b), although some argue that Xi Jinping is striving to become an old-style charismatic leader. China's leaders are "at

Box 3.2
The central-local relationship in China

In China, there are two distinguishing characteristics of the central-local relationship compared with the federal-state relationship in the United States. The first is that the "central government is responsible for policy decisions and the local government is responsible for policy implementation" (中央决策、地方执行) (Liu 2010). In theory, the central government has absolute authority and it can determine any policy for the local government without any limitation. The local government is a subordinate unit of the central government and it cannot oppose policies coming down from higher levels of government in public (Ma, Li, and Ye 2012). To be in compliance with the central government (与中央保持一致) is the most important principle for the local government to observe. Because the central government is focused on policy design and oversight, the scale of the central government is very small in China. In fact, central government officials only account for about 6 percent of the total civil servants in China, compared with about one-third in many developed countries. Similarly, the fiscal expenditure by the central government itself only accounted for 14.6 percent of total government expenditure in 2013, compared with an average of 46 percent for Organization for Economic Cooperation and Development (OECD) countries. In the United Kingdom, United States, and France, this fraction exceeds 50 percent (Lou 2013, 2014). In reality, Chinese local governments have wide discretion over the implementation of policy given China's vast territory, huge population, and the many disparities among different regions.

The second characteristic of the central-local government relationship is to "share powers among the central and subnational governments in same functional policy area, and to maintain a similar structure across the different level of governments" (事权共担, 部门同构) (Zhu and Zhang 2005). In most cases, designation of power and responsibility is not clear among the different levels of government, and no specific law exists to define it. In order to let the central government decide and local government implement, and do it well, the structure of each level of government must be similar because the policy is transmitted level by level through each ministry's parallel body at the subnational levels. For the carbon intensity policy, for example, the NDRC decides on the national target and breaks it down for different provinces. Then, the provincial development and reform commission will receive the target from the central government and break it down for the prefecture city, and so on. So, each level government will need a development and reform commission to complete the policy transmission. Each institution in the central government has a corresponding institution in lower levels of government. The result is that each level of government has a very similar structure.

the top of an inverted funnel that directs the most nettlesome problems up for resolution" (Lampton 2014a, 51). As a result of the system of high centralization, when problems at lower levels cannot be resolved, they are sent upward level by level, so the leaders in Beijing are constantly overloaded with challenges from around the country. To preserve a sense of benign and united resolution among the elites, "consensus has now become the accepted style of decision-making at a national level" (Yu 2014). This apparent consensus masks a highly contentious process that involves a tremendous amount of negotiation among party officials and senior leaders as they try to make decisions (Huang 2014). Although these leaders may be seeking mutual gains—in the language of negotiation theory—they may also find themselves in zero-sum game situations in which painful trade-offs are necessary.

China's governing structure is one requiring both vertical and horizontal coordination. Vertical coordination refers to the fact that the central government hierarchically controls the four other layers of government, with the central government above the provincial government and it in turn above the prefecture, county, and township.

Each level has a similar basic structure compared to the one at the central level. Above the county level, for example, the local party committee has a similar structure to that of the central Standing Committee, and the members of local party committees are usually the most powerful people in the local governments. The secretary of the local party committee is ranked as the number one top leader, and the vice secretary is ranked number two, and he or she would be the governor of the province or mayor of the city. In other words, just like at the central government level, lower levels will have their own congress, government, court, and so forth. In addition, each level of government has almost the same administrative structure: just as there is a National Development and Reform Commission (国家发改委) at the central level, there will be a development and reform commission at the provincial (省发改委) and municipal levels (市发改委) as well. The central Ministry of Environment (国家环保局), for example, is replicated by Environmental Protection Bureaus at the provincial (省环保局) and municipal (地方环保局) levels.

Thus, the organization of the local government is subject to double leadership (双重领导), one from the local government and one from the same organization in the higher level of government, such as the development and reform commission of Zhejiang Province (浙江省发改委), which is led by the government of Zhejiang Province (浙江省政府) and the National Development and Reform Commission (国家发改委) at the same time. In most cases, the leadership from the local government is more important because the local government is responsible for the official positions and money of the local organization directly.

With such a system, negotiations and bargaining are prevalent and necessary not only among the elite, but also among different bureaucracies at different levels. The structure of authority requires that any major policy initiative must gain the active cooperation of many bureaucratic units at multiple levels that are themselves nested in distinct chains of authority. Lieberthal and Oksenberg coined the term fragmented authoritarianism to describe the situation in which consensus building requires bargaining among units at all pertinent levels of the bureaucratic hierarchy, and the policy process is therefore often "protracted, disjointed, and incremental" (Lieberthal and Oksenberg 1988, 22).

For any major new policy or program, at least one or more of the top leaders must enthusiastically support it in order to overcome bureaucratic impasses at lower levels (Lieberthal and Oksenberg 1988, 23). To ultimately convince individuals at the very top level, bureaucrats and party members must go to extraordinary lengths to strive to forge relationships with them. Traditional to Chinese culture is the importance of *guanxi*, the social networks and personal relationships maintained by each individual that are crucial to accomplishing any task. John King Fairbank wrote of the Chinese political tradition that it was a "system of intricate personal relationships that each official had to maintain with his superiors, colleagues, and subordinates" (1979, 116).

Within China, there are two rule systems that work in parallel and sometimes intersect. The first is the formal rule system, with well-established and well-known procedures that produce official documents. The second system is an informal one. It is not public and works through private networks. *Guanxi* is the key element, and in some cases the decisive element, in the informal system, allowing individuals to break through the well-established procedures to accomplish goals. *Guanxi* is usually used for resolving private matters, such as avoiding having to pay a traffic fine. It can also be helpful, however, for obtaining approvals for particular projects or obtaining other private favors.

The value of *guanxi* is largely to break through limitations at each level of government. In other words, in the formal rule system Chinese officials are extremely unlikely to circumvent or go above an official at their level because they must follow the principle of working level by level (逐级). Skipping levels (越级) is forbidden formally, and anyone who tries to do so anyway takes a political risk. But officials can use their *guanxi* in private if they have a relationship with an official at a higher level who could help them accomplish their goal. Within levels, officials will extensively use their formal powers and routine rules to advance or block policy.

Once consensus on a new policy is achieved and official guidance or documents are issued, implementation of the policy relies on both the bureaucratic government institutions and the party at all levels of government. Because the central government institutions are not designed for enforcement, the leaders principally rely on the local CCP and government to implement policy. The means of enforcement are primarily the party-led cadre-evaluation system (干部考核制度) and the government's Local Government Performance Assessment (政府绩效考核) process. In fact, one of the primary roles of the party is to manage cadre affairs, which includes whether or not the cadre is faithfully carrying out the policies and guidance of the CCP.

Every cadre who wishes to be promoted will do his or her best to meet targets set by the higher level of government, promote economic growth, and maintain social stability (Qi et al. 2008). When goals or policies are in conflict, the individual cadre will try to gauge which policies will benefit him maximally and will not choose policies based on the importance set by the higher level of government. Incompatible incentives are very common in Chinese policy documents. The process of implementation in China exhibits the classic principal-agent problem because the local officials always interpret and implement the policies in ways that suit them and decide which ones are in their interests to implement or not, all while they try to convince leaders they have faithfully adhered to their policies (Fewsmith 2013; Miller 2005).

As China shifted to a market-based economy, corruption became rampant because local party officials pursued private interests without suffering consequences so long as they delivered on the higher-level goals of the leaders (Fewsmith 2013). Although Chinese leaders all the way back to Deng Xiaoping have warned about the dangers of corruption, it grew to such an enormous extent that Xi Jinping made it his signature priority to root out corruption. As he said in a speech in January 2013, "We must solidify our resolve, ensure that all cases of corruption are investigated and prosecuted, and that all instances of graft are rectified, continue to remove the breeding grounds for corruption, and further win public trust by making real progress in the fight against corruption" (Xi 2014, 426). Clearly, the people question party legitimacy when corruption runs amok. Or, as Xi stated, "If misconduct is not corrected by being allowed to run rampant, it will build an invisible wall between our Party and the people. As a result, our Party will lose its base, lifeblood, and strength" (ibid.).

Thanks to the influence of the Communist Party, China engages in overall planning. Every five years, the Chinese government releases a new five-year plan that states broad intentions and general guidelines. The plan clarifies priorities and sets vague or more specific goals and targets for the country as a whole and for regions. These plans are how the leadership make clear their goals to every single party cadre and government official in China and create a system of enforcement; leaders of each level and the Chinese government as a whole are subsequently measured by whether or not they meet the goals. The targets for specific regions will not be mentioned in the text of the national five-year plan, but each region will create its own five-year plan based on the national plan and its own specific situation. As a result, there is no one five-year plan but rather a comprehensive system of many five-year plans: industry-by-industry plans, province-by-province, city-by-city, and so forth. In addition, occasionally the government will release a medium- or long-term plan for a priority area, such as the National Medium- and Long-Term Plan for Science and Technology Development, a fifteen-year plan issued in 2006 that set a goal of spending 2.6 percent of GDP on research and development and reducing the country's reliance on foreign technology to below 30 percent (Xinhua 2006).

The rigor of the planning process varies, and in many cases it is based not on formal analysis or mathematical calculations but rather primarily on political factors. One limitation is the lack of government personnel. The Chinese government is lean relative to the United States. Although it is difficult to make direct comparisons, as of the end of 2015 China had seven million formal civil servants, only twice as many as in the United States, even though its population is more than four times larger (Lu and Cheng 2016). In addition, there are many other state-financed employees that are not civil servants, estimated at about fifty million (Chi 2016). The other problem with the planning process is that there is a tendency to set goals that are easy to accomplish so that understated targets can be overachieved.

Several other factors significantly influence the policy process, including the training and background of Chinese leaders. In the past, most Chinese government officials were trained as scientists and engineers, not as lawyers or in public administration. Increasingly, government officials have social science backgrounds. In the previous Politburo Standing Committees during the terms of former presidents Jiang Zemin (江泽民) and Hu Jintao (胡锦涛), for example, most of Standing Committee members were trained as scientists and engineers. But in the last Politburo Standing Committee, some were trained in law, such as Li Keqiang (李克强); some in economics, such as Xi Jinping (习近平), Zhang Gaoli (张高丽), and Zhang Dejiang (张德江); and Wang Qishan (王岐山) was trained in history. Currently, the top leaders almost all have economics and law degrees. In economics, there is Xi Jinping, Li Zhanshu (栗战书), Wang Yang (汪洋), Zhao Leji (赵乐际), and Han Zheng (韩正). In law, there is Li Keqiang (李克强) and Wang Huning (王沪宁). But their educational backgrounds are not as important as they were for their predecessors, because in the Jiang and Hu eras, most of the top leaders finished their education before entering the bureaucracy. In the Xi period, most senior officials receive their degrees mid-career, and therefore their degrees serve more symbolic roles.

Perhaps because there is a strong respect for technical expertise, and perhaps because of the limited number of overburdened officials in the central government, government agencies actively support and seek the advice of outside experts. Formal advisory bodies are often established to help government officials contemplate and evaluate policies. Each ministry and department often heavily relies upon one or more research institutes. Outside experts also influence the planning process. More than four thousand experts from various fields contributed to the formulation of the medium- and long-term plan on science and technology, for example (Zhou 2014).

Academics or technical experts must cultivate their own *guanxi* to be influential in policy processes. There is a distinct advantage to being within close geographic proximity to the government officials one wants to reach. Also, of course, many of the research institutions affiliated with the central government are based in Beijing. For central government policy, Beijing-based academics tend to have considerably more influence than those far from the seat of government. Prior connections and networks can and do exist, however, and new ones can be established by determined individual experts.

A major difference in China's contemporary policy process relative to the post–Cultural Revolution era is the growth in influence of corporate and other private-sector interests. Gradually but steadily, the Chinese government has encouraged entrepreneurship that contributes to economic growth. The transition to a market-oriented economy has been called "capitalism with Chinese characteristics" (Huang 2008), and it has created enormous wealth for some individuals, many of whom are connected to the government in some way. This system continues to incentivize corruption because governmental officials within the system can identify opportunities for themselves and their friends and family that ordinary citizens cannot see. The highest-level official to be convicted of corruption was Zhou Yongkang (周永康), a former member of the Politburo Standing Committee and the head of China's security services, although he may have been prosecuted in part for political reasons. When he received his sentence of life in prison, the estimated value of his family's assets was USD 14 billion or RMB 90 billion (Wu 2015).

Meanwhile, powerful state-owned enterprises in strategic industries such as banking, energy, telecommunication, and aviation still exist, led by senior CCP party members. These SOEs can be enormously influential in policy processes because they are both "inside" and "outside" the government. The top leaders of major state-owned enterprises can advocate in their industries' self-interests as government officials with ministerial ranks. The Central Organization Department of the Central Committee of the CPC (中组部) manages the top SOE leaders with input from the state-owned Assets Supervision and Administration Commission, the SOE watchdog. These corporate leaders thus have incentive to place political or policy considerations over corporate efficiency, but on the other hand, these state-owned enterprises are known for effectively resisting reform because the status quo serves them well. Sometimes these firms are forced to sacrifice for the public interest, such as when China's state-owned oil companies were forced to buy oil at high prices on the global market during the 2000s and sell at low prices domestically to maintain stability and economic growth within China (Stocking and Dinan 2015). Corporate and industrial interest groups are known to influence governmental decision-making processes, either through fostering government policy gridlock or altering policies in their own favor.

The contribution of *civil society* to policy process in China is another area where change has occurred. Some interest groups have emerged that try to express their interests, and certain types of NGOs are now allowed. Industry associations, in particular, have been encouraged ever since the early 1980s in part because they generally are thought to support economic reform and because they have an interest in a predictable business environment (Fewsmith 2013). Other civil society organizations that are encouraged are those contributing to social good, such as environmental education. Although civil society interest groups appeared to grow more organically after the turn of the century, Xi Jinping and other conservative leaders have succeeded in reining them in, issuing an enormously controversial policy requiring all foreign civil society organizations to have a sponsoring government agency, which in turn has responsibility for supervising them. This relationship has been unfavorably compared to the one between a newly married individual and his or her mother-in-law. Domestically, civil society organizations are subject not only to the relevant Ministry of Civil Affairs (民政部), which is responsible for the management of civil society organizations, but also to the functional ministry in their area. Civil society organizations are allowed to operate only within their designated localities and engage only in the activities approved when they register (Yu 2011, 79).

On the whole, however, civil society organizations in China appear to have very little influence on the policy process, though civil society finds ways to express its preferences and displeasures. The most potent examples of civil society pressure coming to bear on central government decisions are in the areas of health and environment. For example, an enormous baby milk contamination scandal exploded in 2008. Shady companies that were cutting costs mixed the chemical additive melamine with powdered milk and infant formula in order to boost volume and the apparent protein content. As infants sickened and died around China, Chinese parents and grandparents grew increasingly angry about food safety. The Chinese government subsequently passed new food-safety regulations, prosecuted the firms that were at fault, and created a food-safety commission to address the widespread concerns about food contamination in China (Bottemiller 2010; State Council 2010). Food safety is still an ongoing concern for many Chinese families.

Similarly, the severe urban air pollution in Beijing and other cities in China gradually became a source of massive discontent, with urban residents being forced for the first times in their lives to wear face masks whenever they go outdoors, their children developing asthma, bronchitis, and other respiratory ailments, and schools and other activities being canceled on "red alert" days. Both the central and municipal governments responded to the air pollution problem, passing new regulations, restricting coal consumption and vehicles in certain areas, and occasionally temporarily banning economic activity altogether to clear the skies, but the efforts have not been successful at eliminating hazardous air pollution.

In both of these examples, the government is responding reactively to civil society concerns, but not because of pressure from civil society organizations. Many Chinese still have major concerns about the safety of their food and the quality of their air. The problem is that the regulatory system is not transparent, adequately resourced, or institutionally coherent. In the milk case, the government had exempted some firms from inspections, failed to recall milk products even when the problems were known in certain provinces, and regulated domestic products differently from those destined for foreign markets (Hu 2008). Although there are provisions in Article 35 of the Legislation Law for public input into the policy process, and the public is invited to submit opinions about many draft policies, it is unclear how much influence these public comments actually have on policy decisions.

Compared with the situation in the United States, Chinese interest groups have some special attributes. The most important one is that Chinese interest groups do not plan to be an independent force or voice, which means that they lack the legitimacy of US interest groups.[7] To a large extent, interest group is a negative term in the Chinese culture because such groups are perceived to work only on behalf of their own interests, without the public interest at heart. For example, when the NDRC designed a policy to form a new national natural gas pipeline company that would include pipelines mainly from PetroChina, PetroChina's objection was not that the company's own interests would be harmed, but rather that the national interest would be harmed because there would not be enough investment to build and maintain gas pipelines in the country and therefore natural gas development would not achieve the national goal. Although the argument was true to some degree, the example illustrates how public interest arguments can be used for private purposes.[8] Interest groups in China also can hide others behind them, disguising their true identities, or can mask private concerns as public ones.

The judiciary in China is less important relative to other organs in the governmen, and also less independent than in the United States. The judiciary system, the procuratorate system (检察院), and the police system (公安) are considered public security and justice organs (公检法系统) in China. On the party side, there is a Political and Legal Affairs Commission (政法委) that leads and coordinates the organs of the public security, the procuracy, and the court. At the central level is the Political and Legal Affairs Commission of the Central Committee (中央政法委). The previous leader of this commission is Zhou Yongkang (周永康), and he is a former member of the Politburo Standing Committee. The current leader is Guo Shenkun (郭声琨), who is of a member of the Political Bureau (政治局委员; twenty-five persons). The leader of the Supreme People's Court, Zhou Qiang (周强), has weaker standing because he is only a member of the two-hundred-person Central Committee (中央委员); he is not a member of the Politburo Standing Committee (常委) or even a member of the Political Bureau (政治局委员). His inferior rank in the party system reflects the status of the judiciary system in China (SPC 2017a).

Second, the independence of China's judiciary cannot be ensured. At the local government level, a structure exists that is similar to the central government structure. Usually, the local government provides the local people's court's budget. The local government has the most important influence on the appointments in the local courts, so the independence of a local judiciary system is hard to achieve. One of the current reforms in the judiciary system is to strengthen the independence of the local judiciary system by unifying the management of staff, funds, and properties of the courts below the level of the province.

Advantages and Disadvantages of the US and Chinese Policymaking Systems

Now that we have described the structures, actors, processes, and approaches to policy in the United States and China, it is possible to consider some of the advantages and disadvantages of each system. Here we aim to illuminate some of the key differences, rather than elucidate all the possible answers. As illustrated in table 3.7, there are substantial advantages to each system, but neither is without considerable disadvantages too.

Table 3.7
Advantages and disadvantages of US and Chinese policymaking systems

United States		China	
Advantages	*Disadvantages*	*Advantages*	*Disadvantages*
Checks and balances	Blocking power of individual branches of government; gridlock	Unifying force of the CCP to create national campaigns and concentrate resources	Lack of consensus can create paralysis; no checks and balances on leaders
Implementation and enforcement through federal and state regulatory institutions	Explicit influence on policies through campaign contributions	Flexibility in implementation and enforcement for local officials	Weak ability to enforce central government policies, corruption
Public and private sector have ample formal and informal opportunities for influencing policy	Long-term broader public interest can be lost due to short-term election pressures	Central government has ability to take the longer view, consider national interest, unify country	Public has few means for influencing policy process
Federal and state-level policy process allow for policy processes to move forward in parallel	Regulatory environment can be complex with fifty states plus federal government	Central government can create and issue new policies quickly if consensus exists	Implementation of central government policies often is resisted or ignored by local government officials
Most government officials are open to briefings by technical experts	Technical expert input not usually solicited; Congressional Office of Technology Assessment abolished	Technical input can be high, is often solicited	Technical experts with better access to government officials are more influential
Dissent and opposition can be vocally and visibly raised	Monied interests can influence public perceptions of issues and foster confusion, muddy politics of issues	Central government can be more decisive due to lack of open dissent	Private criticism may be welcomed, but public criticism can cause marginalization or lack of promotion; consistency with policy view is important for advancement
		Conduct five-year and medium-term planning cycles	

For advocates of action on climate change in the United States, the US system of checks and balances has been some consolation during recent Republican presidencies because either the US Congress or the Supreme Court provides avenues for blocking policies counterproductive to climate mitigation and even possible avenues for promoting new policies. On the other hand, each individual branch has considerable blocking power, and when at least two are in a blocking position it can be extremely difficult to advance new policy. Still, due to checks and balances, one or more of the branches of the US government will stop most truly harmful or unjust policies from being implemented.

The second advantage of the US system is the federalist system of government, which permits the federal government to develop and implement policies that reach all the way to the local level. In addition, the US system can be characterized as one with multilevel governance (Selin and VanDeveer 2009), which means that states, counties, cities, and towns can all develop and implement their own policies, without the need for approval from the federal government. The freedom to pursue state and local policies allows them to develop policies suitable to the unique circumstances of their jurisdictions, as well as experiment as leaders in their own right with new policy approaches. The boundaries between federal and state authorities are fairly clear, and there is legal recourse in cases of dispute.

Because of politics, the strength of interest groups, and the separation of powers, it can take a very long time to make progress on important issues, and sometimes such issues cannot be resolved at all. This inherent slowness, while conceived by the framers of the Constitution to be an advantage in a deliberative democracy, can also be considered a major weakness because the US government frequently fails to address problems in a timely way; indeed, it sometimes loses the opportunity to solve a problem in time. Climate change policy is a case in point.

Because elected leaders only serve for a certain period of time in the United States, they have a tendency to put much more emphasis on present or short-term challenges and opportunities. Relatedly, because elected leaders are beholden to the people who elected them, they will more likely focus on pleasing narrow constituencies rather than the good of the entire nation. Naturally, it is also difficult for them to consider the longer-term consequences of their actions unless they are confident of reelection. A

Chinese-style medium- or long-term plan is almost impossible to imagine in the US context.

By contrast, the CCP provides a unifying force with the ability to create national campaigns and concentrate resources around priorities. On the other hand, if there is no consensus within the party, it is extremely difficult to advance policy. In addition, there are few checks and balances in the Chinese policymaking system because the most important characteristic of China's political regime is centralization (集权). The first advantage of centralization is that policy can be decided in a very short time if necessary because the decision-making power is concentrated in the hands of a small number of top leaders, especially at the central government level. The CCP plays a decisive role in managing and coordinating different centers of power and interest groups through their rigorous organization in China. In fact, the general secretary of the CCP has the absolute power to make any decision in any important area given how few checks and balances there are in the regime.[9]

The second advantage to the centralization of China's policymaking system is that the government can make policy for the good of the whole country without being unduly influenced by any one region or stakeholder. In addition, the government can allocate considerable resources behind national goals, and it can do so quickly. With this type of regime, sensible and virtuous leaders are desirable.

The first disadvantage of China's policymaking system is that it is impossible to ensure that good leaders are in place to make wise policy decisions all the time. The "bad emperor problem" is the Achilles' heel of any centralized political system. According the Chinese history, there have been far fewer good emperors compared with bad ones, and even the good emperors did not always make shrewd policy decisions. Although modern China is very different from ancient China in so many areas, the mode of the political power allocation has *not* changed too much, which means only a few people can grasp political power—and the highest person has the absolute power. This challenge persists in modern China. Because decision-making power is held by so few leaders, the they have enormous influence over their region of responsibility, whether it be the party general secretary for the entire country, a provincial party secretary for his province, or a municipal party secretary for her city. If government leaders make a big mistake, it is hard to correct it; Mao's decision to launch the Cultural Revolution is a case in point.

The second disadvantage of China's policymaking system is its serious principal-agent problem, which results in an inability to ensure the intended outcome of a policy. Because the central government is responsible for policy formulation and decisions and the local government is responsible for policy implementation, the central government loses control over policy outcomes. The central government does not have the same incentives as lower levels of government, and an information asymmetry exists, which is classic to principal-agent theory. The local government always has its own interests and always has more information about the real situation on the ground, but it lacks incentives to share this information. Although central government can and does use performance evaluations and frequent inspection tours to alleviate this challenge, it remains a fundamental problem with the Chinese system.

4 National Target Formation

How do countries set climate change policy goals in the first place? This chapter explores the goal-setting processes in China and the United States and clarifies the similarities and differences in their policymaking processes.

In the historic joint announcement of President Xi Jinping and President Barack Obama, the United States stated that it would achieve a 26–28 percent reduction in greenhouse gas emissions below 2005 levels by 2025. For its part, China declared that its carbon emissions would peak around 2030, at which time nonfossil energy sources would account for 20 percent of its energy supply. In a reciprocal nod, the United States pledged to make "best efforts" to achieve the upper end of its emissions-reduction range, and China likewise said it would make "best efforts" to peak earlier than 2030.

The Targets and Timetables Approach

To begin this chapter, we must understand why both countries are using the so-called targets and timetables approach—that is, specifying how much GHG emissions will be reduced by a certain date. This approach originally was used in the 1987 Montreal Protocol on Substances that Deplete the Ozone Layer, when reductions in stratospheric ozone-depleting chlorofluorocarbons (CFCs) were also specified in precisely this manner. Five years later, when the original UN Framework Convention on Climate Change (UNFCCC) was adopted at the Earth Summit in Rio de Janeiro in 1992, the negotiators utilized the same approach. In the 1992 UNFCCC, industrialized countries agreed to aim to reduce their emissions to 1990 levels on a voluntary basis. In the subsequent agreement, known as the Kyoto Protocol to the UNFCCC, negotiated in 1997, industrialized countries as a whole committed to reduce their emissions 5 percent below existing 1990 levels in the commitment period 2008–2012, and individual country commitments known as Quantified Emission Limitation or Reduction Commitments (QELROs) were spelled out in an appendix of that agreement (UNFCCC 1998).

In the 2015 Paris Agreement, countries utilized a new, bottom-up approach: instead of negotiating these targets and timetables as a whole, each country determined the target by itself. This concept has become known as *pledge and review* because the idea is that each country makes a pledge, which is then reviewed for adequacy in the context of the international UNFCCC negotiations. Prior to Paris, and beginning with the United States and China, nearly every country declared these pledges, technically known as *intended nationally determined contributions* (INDCs). These INDCs became *nationally determined contributions* (NDCs) when the Paris Agreement entered into force late in 2016.[1]

The targets and timetables approach has not been seriously questioned as an approach by any country, although European governments at one time proposed harmonizing *policies and measures* (PAMS) as an alternative approach. This approach might have manifested itself as performance standards or emissions taxes that were consistent across countries. To provide two examples, countries could have agreed to enact the same fuel-economy standards for motor vehicles or to impose the same levels of carbon taxes in their economies to level the playing field economically.

A number of proposals over the years have also been made to allocate emissions quotas based on equity considerations, such as equal per capita emissions across countries, but these proposals never gained real momentum in the international climate negotiations.[2]

Emission-reduction targets can be expressed in many forms, such as a percentage reduction in absolute emissions—of all greenhouse gases or of just one gas, such as CO_2. Such targets may include emissions from land-use change—for example, carbon emissions from deforestation. Switzerland currently has the most ambitious emissions-reduction target of any industrialized country: a 50 percent reduction below 1990 levels by 2030.

China pioneered the approach of *peaking targets*, in which developing countries clarify a time by which their emissions will no longer rise, implying that the emissions will begin to be reduced thereafter—but of course there could be an extended plateau at the level of the peak. After China announced its 2030 peak, Mexico established a more ambitious timetable by specifying a 2026 peak.

A third approach is to declare an *intensity target*, which is the amount of GHG emissions per unit of economic output (e.g., GHG emissions/GDP). Intensity targets are chosen when a country wishes to avoid constraints on economic growth but thinks that it can achieve reductions through technical improvements or switching to cleaner fuels that would result in a less carbon-intensive economy. Building wind farms and shutting down the equivalent electricity capacity in coal-fired power plants would reduce the carbon intensity of an economy, for example. China announced an additional carbon-intensity goal of 60–65 percent below 2005 levels by 2030 as part of its formal submission of its INDC to the UNFCCC.

Some countries have specified their targets as reductions from business as usual (BAU). In such cases, the countries usually (but not always) clarify what they expect their BAU emissions would look like over a certain period and then how much they expect to reduce emissions below this trajectory. BAU targets are usually *growth targets*; in other words, emissions are expected to grow, just at a lower rate than they otherwise would have done. Ethiopia, for example, set forth a target of limiting GHGs including those from land-use change to 145 MT CO_2eq by 2030. This represents a reduction of 64 percent below the Ethiopian BAU scenario by 2030, for which net emissions are projected to reach 400 MT CO_2eq. The challenge with growth targets is that a high BAU trajectory can be claimed by a country to make its emissions-reduction target seem more impressive.

The bottom line is that few targets are easily comparable given all these alternative approaches. It is almost impossible to make apples-to-apples comparisons of relative effort. One method is to consider the annual *rates* of reductions in emissions or emissions intensity (if emissions are still growing).[3]

Domestic Approaches to Setting Emission-Reduction Targets

US Target Setting

In the United States, the setting of emissions-reduction targets has historically been handled by the executive branch of the US government through a process coordinated by the White House. Federal agencies provide input into the process that ultimately culminates with a final decision made by the president. The US Congress has the power to set domestically binding emission-reduction targets for the United States through climate change legislation, but it never has been able to pass a climate change bill.

The first US target was set in the context of the UNFCCC itself, although it can hardly be remembered as a "target." During the UNFCCC negotiations concluding in 1992, the United States needed to reach consensus with other industrialized countries because the same target was contemplated for all of them. International forces thus drove the domestic process, and the State Department coordinated it in consultation with the White House. At the Earth Summit in Rio, the ultimate objective of the convention was agreed upon, namely to "prevent dangerous anthropogenic interference with the climate system" (UNFCCC 2017a). As for the specific target, negotiators began to narrow down to a target of reducing emissions to 1990 levels by the year 2000 in the context of Articles 4(a) and 4(b). According to Daniel Reifsnyder (2017), a member of the US delegation to the UNFCCC, "targets and timetables were a complete red-line for the George H. W. Bush administration."

A key disagreement in Rio was whether or not the industrialized countries "shall" achieve the 1990 level target or would "aim" to achieve the target. The United States insisted on the word *aim*, which rendered the target completely voluntary, contrary to the wishes of many other countries (Gupta 2010). The year 2000 was not specified in the end, either. Rather, Article 4(a) mentions "a return by the end of the present decade to earlier levels of anthropogenic emissions," which could have been interpreted to mean the end of either 1999 or 2000. It was also not specified whether or not industrialized countries needed to remain at that level after the year 2000. At the time, the "aim" was interpreted by everyone in the US government as completely voluntary for the United States, so the US Senate quickly ratified the UNFCCC.

Several years later, the UNFCCC negotiators established the Berlin Mandate, which called for a more ambitious, legally binding treaty to be negotiated no later than 1997. The resulting agreement was the Kyoto Protocol, which indeed set binding commitments for industrialized countries (Annex I) for the period 2008–2012. The US target within the Kyoto Protocol was a 7 percent reduction below 1990 levels by 2012. President Bill Clinton ultimately approved this target with active encouragement from his vice president, Al Gore, who personally traveled to Kyoto, Japan, to demonstrate the depth of US support for the negotiations and help reach a final deal. The US Kyoto target was not established through a detailed domestic policy process in the United States but rather was a target arrived at through international negotiation.

The Kyoto Protocol was not submitted to, nor ratified by, the US Senate due to overwhelming opposition in the same because of the lack of binding targets for major developing countries, most specifically China and India. Months before the final Kyoto negotiations, two US senators, Republican Chuck Hagel and Democrat Robert Byrd, sponsored a nonbinding resolution that won unanimous approval from the rest of the Senate. Resolutions are not formal legislation, but in this case the resolution expressed the "sense" of the Senate and indicated how senators were likely to vote on treaty ratification. This resolution specified that the United States should not be a signatory to any agreement that would "mandate new commitments to limit or reduce greenhouse gas emissions for the Annex I Parties, unless the protocol or other agreement also mandates new specific scheduled commitments to limit or reduce greenhouse gas emissions for Developing Country Parties within the same compliance period" (Byrd-Hagel Resolution 1997). In a later interview, Senator Hagel commented that he did not perceive the Kyoto Protocol to be equitable because major countries like China did not have the same commitments as the United States, despite their vastly different levels of development. He said, "I did think it was not fair, because it did not include all the nations of the world. ... Only 30 nations of the world, essentially, were given mandates as to roll back their man-made GHG emissions to some 5 to 7 percent below those 1990 levels. But it did not include nations like China, South Korea or India. Most nations of the world were left out. So how was that fair?" (PBS Frontline 2007).

In the next presidential election, climate change policy advocate Vice President Gore lost the Electoral College but not the popular vote, so it was up to incoming Republican President George W. Bush to determine the US climate change target under his administration given the lack of US ratification of the Kyoto Protocol. In February 2002, President Bush announced that the United States would, on a voluntary basis, reduce the intensity of GHGs by 18 percent below 2002 levels by 2012, the year when the first commitment period of the Kyoto Protocol would end. Analysis at the time projected that this intensity target would cause emissions to rise approximately 12 percent over that time period (C2ES 2002). Actually, US emissions fell from 7,185 MMT CO_2eq in 2002 to 6,643 MMT CO_2eq in 2012, which represented a 7.5 percent reduction below 2002 levels and 4 percent above 1990 levels (recall that the Kyoto target for the United States was a 7 percent reduction *below* 1990 levels by 2012; EPA 2018). In retrospect, the Bush administration failed to take into account the shifts in the electricity markets away from carbon-intensive coal and toward natural gas.

Observers of the Bush administration's climate policy conclude that industry groups exerted substantial influence during this period. In documents obtained through the Freedom of Information Act (FOIA), State Department officials were found "thanking Exxon executives for the company's 'active involvement' in helping to determine climate change policy, and also seeking its advice on what climate change policies the company might find acceptable," according to an expose by the *Guardian* (Vidal 2005). A *New York Times* exposé found that eighteen "of the energy industry's top 25 financial contributors to the Republican Party advised Vice President Dick Cheney's national energy task force" in 2001 (Van Natta and Banerjee 2002).

Although the United States never ratified the Kyoto Protocol, it remained a party to the UNFCCC and thus continued to participate in the annual international Conference of Parties. To develop a successor to the Kyoto Protocol, a new deadline was set by UNFCCC negotiators of December 2009 at negotiations scheduled to take place in Copenhagen, Denmark.

President Barack Obama was elected in 2008, and he immediately pledged "a new chapter" in climate policy in a YouTube speech; specifically, he said that the United States should establish a cap-and-trade program to reduce GHG emissions to 1990 levels by 2020 and further reduce them 80 percent by 2050 (Obama 2008). In 2009, President Obama challenged Congress to establish such a cap-and-trade program in a special address to Congress (Samuelson 2009). Congressmen Henry Waxman and Edward Markey took up the challenge and developed a new bill during the 111th Congress known as the American Clean Energy and Security Act (the Waxman-Markey bill for short). This bill, HR 2454, established a new target of reducing US GHG emissions 17 percent below 2005 levels by 2020 and established a domestic cap-and-trade program to do so. This bill passed the US House of Representatives 219 to 212 in June 2009, but Senate Majority Leader Harry Reid never brought its equivalent to the floor in the Senate, led by Senators John Kerry, Joseph Lieberman, and Lindsey Graham, for a vote. Although he was supportive, President Obama failed to throw all of his weight behind the bill, reportedly out of political concerns, a decision to prioritize health care reform, and the unfortunate timing of the Macondo well blowout in the Gulf of Mexico (unfortunate because the Senate bill expanded offshore oil drilling as a trade-off for industry; Lizza 2010). Even though the Democratic Party held the majority at the time, too many Democratic senators from coal mining and manufacturing states opposed the bill.

Nonetheless, when President Obama traveled to the Copenhagen meeting in December 2009, the Senate bill had not yet failed. Obama adopted the Waxman-Markey bill target of 17 percent below 2005 levels by 2020 and made it an official pledge of the US government. Although the Copenhagen meeting did not result in a formal agreement as planned, President Obama decided to develop a plan to honor the target anyway. In June 2013, he announced a Climate Action Plan with three components: "cutting carbon pollution" in the United States (mitigation), preparing the United States for the impacts of climate change (adaptation/resilience), and international leadership on climate (White House 2013). The mitigation portion of the Climate Action Plan was intended to set the United States on course to achieving the 2020 target through a suite of domestic policies and programs.

Meanwhile, the international community set a new deadline for the international negotiations under the UNFCCC: December 2015. A new process was envisioned in order to avoid Kyoto and Copenhagen-like failures to achieve a globally acceptable agreement. In the new process, which would ultimately result in the Paris Agreement on Climate Change, a "bottom-up" approach would be used, in which countries would put forth their INDCs in advance of the December 2015 meeting. In other words, targets would not be negotiated by countries but rather be self-generated, as discussed at the beginning of this chapter. The Obama administration strongly promoted and supported this approach because senior officials led by the special envoy for climate change, Todd Stern, believed it could lead to a universal agreement in which all countries participated, which could break down the historic barriers between industrialized and developing countries (Stern 2015).

To prepare a strong and well-supported INDC, White House staff led by Deputy Director for Energy and Climate Rick Duke initiated an interagency process in 2014 to identify policies and measures that would lead to a more ambitious post-2020 target for the United States, taking into account the policies already implemented during President Obama's first term—most significantly the new fuel-economy standards for automobiles but also including the loan guarantee program administered by the US Department of Energy, investments in energy RD&D, existing production and investment tax credits for renewable energy, and energy-efficiency standards issued by DOE. In addition to these existing policies, new and additional potential policies and regulations were contemplated, including the Clean Power Plan, a regulatory rule-making that was later finalized by the EPA. All of these policies were modeled using the EPSA-NEMS integrated energy-assessment model (US Department of State 2016). Aside from the Clean Power Plan, other major new policies included EPA's methane rule, new regulations to limit HFCs under the Significant New Alternatives Policy (SNAP) program and as obligated by the Montreal Protocol, and heavy-duty-vehicle efficiency standards. Considerable uncertainty about land-use change and forestry emissions prevailed, but this sector was ultimately included in the INDC.

The Domestic Policy Council (DPC) and the Council on Environmental Quality (CEQ) coordinated the interagency process in the White House under the watchful eye of the counselor to the president who was ultimately responsible for climate policy at the time, John Podesta. Both the DPC and CEQ play coordinating roles in the federal government because they are situated in the Executive Office of the President. Somewhat unusual was the special role of the counselor to the president because he had been brought into the White House to focus on implementation of a few of President Obama's highest priorities, including domestic climate policy. The interagency process was a slow, iterative, bottom-up process that included all the relevant federal government agencies and other offices of the White House, including the National Security Council and the Office of Science and Technology Policy. As is normally the case, the process started at the staff level and gradually rose for a recommendation by cabinet-level officials to the president.

Meanwhile, in late 2013, the Obama administration began to consider the possibility of jointly announcing INDCs together with China to catalyze other countries to come forward early with their INDCs. After proposing the idea of jointly announcing the INDCs to the Chinese government in early 2014, the Obama administration had to match its internal decision-making process to the schedule for the US-China bilateral negotiations. The Obama-Xi Presidential Summit scheduled for November 2014 imposed an internal deadline on the US INDC decision-making process.

With the election of Donald Trump as president, the feasibility of implementing the US target was again imperiled, strikingly similarly to the period after Al Gore lost to George W. Bush. In the case of President Trump, he campaigned on a pledge to withdraw from the Paris Agreement, as well as to remove some of the signature regulatory policies that would have enabled the United States to meet its pledge, most significantly the Clean Power Plan. More discussion on the implementation of the US target follows in chapter 5, including how the Trump administration took action in its first year.

Chinese Target Setting

In the history of national climate target formation policy in China, four distinct stages can be observed: (1) a "no compulsive target" period, (2) adding an energy-intensity target, (3) adding a carbon-intensity target, and, finally, (4) establishing an emissions peak. These stages occurred sequentially, and each coincided with a new five-year plan in China. The five-year planning process dates back to the Communist Revolution in Soviet Union that produced the first five-year plan for 1928–1932, after which China's leaders adopted the Soviet method of centralized planning for the economy when Communist China was founded in 1949. The first five-year plan in China was issued for 1953–1957, and the thirteenth five-year plan was issued for 2016–2020. In the early years, when China was a centrally planned economy, the five-year plan played a decisive role because it would directly allocate the resources of the whole nation. When China transitioned to a market economy after the reform, the five-year plan played more of a guideline role, emphasizing the direction and targets of the national economy. Therefore, before the tenth five-year plan, five-year plans were called 计划 (jihua); but since the eleventh five-year plan, they have been called 规划 (guihua). Although in English they appear to be the same, in Chinese, the connotation is very different.[4] Because of the long practice of establishing national social and economic targets as part of the five-year plans, the concept of establishing climate-related emissions targets was perfectly acceptable in China and in fact was embraced after 2005.

China was also involved in both the Montreal Protocol and UNFCCC negotiations from the start and thus had familiarity with the target-based approach embraced by these conventions. One important difference from the United States is that, as a developing country, China did not have climate targets under the original 1992 UNFCCC convention, nor even in the 1997 Kyoto Protocol, although it could establish them on a voluntary basis. China's first formal targets were proposed in Copenhagen in 2009, but that summit failed to achieve a new treaty. Later, China established the set of targets described in chapter 1 in the context of the US-China agreement and later enshrined in the international Paris Agreement on Climate Change of 2015.

Prior to 2005 can be considered the "voluntary target" period. This was the year that the tenth five-year plan (2001–2005) ended. During the tenth five-year plan, there were no explicit compulsory targets in the energy area and not even any energy-conservation targets. Vague statements, such as "make significant progress in resource-saving and protection," were included. Still, the Chinese government was aware of the climate change issue, primarily due to its engagement in international negotiations. In 1992, the National People's Congress approved China's accession to the UNFCCC, and the Chinese government also ratified the Kyoto Protocol in 1997.

During the 1990s, the Chinese government adopted a strategy of addressing climate change within the framework of sustainable development. To do so, a National Coordination Group on Climate Change was established in 1990 in support of the international negotiations on climate change (NDRC 2012). Another leading group was established that was cochaired by a deputy minister of the State Science and Technology Commission and a deputy minister of the State Planning Commission to "organize and coordinate the formulation and implementation of China's Agenda 21" (United Nations 1997). "China's Agenda 21: A White Paper on Population, Environment and Development in the 21st Century" was published in 1994. That same year, the State Council also issued a directive to government agencies to consider China's Agenda 21 as a strategic guideline for the next five-year plan (1996–2000) (UN DESA 2017).

In early years, the climate issue in China was considered to be mainly an international and scientific problem, and it had little relationship with domestic policy (Lewis 2008). The Ministry of Foreign Affairs and

China Meteorological Administration (气象局) played the most important roles during that period. The former was in charge of international issues, and the latter was in charge of climate science at the time. In 1990, the National Coordination Group on Climate Change (国家气候变化协调小组) was founded, and its office was based in the China Meteorological Administration.

As time passed, the climate issue became increasingly entwined with domestic policy, especially in the energy and economic sectors in China. The State Development Planning Commission (predecessor to the NDRC) played an increasingly important role. In 1998, a National Strategy and Coordination Group on Climate Change (国家气候变化对策协调小组) was founded, which was the successor to the National Coordination Group on Climate Change, but its office moved to the State Planning Commission; the top leader of the group was Zeng Peiyan (曾培炎), who was the minister of the SPC at the time.

After China ratified the Kyoto Protocol in August 2002, the Chinese government reconstituted the National Strategy and Coordination Group on Climate Change (国家气候对策协调小组) and it was enlarged to include seventeen ministries. Its office was still based in the NDRC. Aside from the NDRC, which led the group, the Ministry of Foreign Affairs (外交部), Ministry of Science and Technology (科技部), China Meteorological Administration (气象局), and State Environmental Protection Administration (环保总局) were appointed as vice-leader institutions (副组长单位). The top leader of the group was Ma Kai (马凯), who was the minister of the NDRC at the time.

In 2007, when the National Leading Group to Address Climate Change (国家应对气候变化领导小组) was founded, the top leader of the group changed to Premier Wen Jiabao (温家宝). When Wen retired in 2012, the top leader of the group changed to the new premier, Li Keqiang (李克强), and its office was still located within the NDRC. In 2008, the division on climate change (应对气候变化司) within the NDRC was founded, and Su Wei (苏伟) was transferred from the Ministry of Foreign Affairs to become the director of this department. He was also a director the office of the National Leading Group to Address Climate Change (国家应对气候变化领导小组协调联络办公室秘书长). In summary, the office of the National Leading Group to Address Climate Change and the climate change department under the NDRC were one and the same (一个机构,两个牌子). See table 4.1 for the ministerial composition of the National Leading Group (NDRC 2017b).

Table 4.1

China's leading group on climate change

National

National Development & Reform Commission (国家发改委)	Ministry of Health (卫生部)
Ministry of Finance (财政部)	National Bureau of Statistics (统计局)
Ministry of Foreign Affairs (外交部)	State Forestry Administration (林业局)
Ministry of Science & Technology (科技部)	China Meteorological Administration (气象局)
Ministry of Industry & Information Technology (工信部)	National Energy Administration (能源局)
Ministry of Land and Resources (国土部)	China Civil Aviation Administration (民航局)
Ministry of Environmental Protection (环保部)	State Oceanic Administration (海洋局)
Ministry of Housing and Urban-Rural Development (住建部)	
Ministry of Transport (交通部)	
Ministry of Water Resources (水利部)	

Provincial

Ministry of Agriculture (农业部)	Provincial Development & Reform Commission
Ministry of Commerce (商务部)	Provincial Finance Department

Officially, the National Development and Reform Commission (NDRC; 国家发改委) was responsible for climate change policy until 2018, at least as it related to CO_2 and methane (CH_4). In 2018, the government announced a reform and created a new Ministry of Ecology and Environment (MEE) (生态环境部), replacing the Ministry of Environmental Protection and moving the responsibility for climate change policy from the NDRC to the new ministry (Xinhua 2018). The division on climate change (气候司) will move to the new MEE, as will the specific work of the National Leading Group to Address Climate Change (国家应对气候变化领导小组). The National Nuclear Safety Administration is also administered by MEE. The NDRC retains responsibility for China's industrial and energy policies.

The NDRC's Climate Change Division (应对气候变化司) issued many climate-related policies over the years—for example, establishing Carbon Emissions Trading Pilots and Low-Carbon Pilot Cities, to name just two. The National Energy Administration (国家能源局) also falls under the

NDRC's authority, and it is particularly responsible for formulating policies related to energy, many of which have a direct bearing on climate, such as those on optimization of energy mixes and the development of renewable energy.

The former Ministry of Environmental Protection (MEP; 环保部) always had authority for conventional pollution, and thus many of the pollution-reduction measures it implemented had cobenefits for greenhouse gas emissions. MEP also had authority over ozone-depleting chemicals, some of which are also heat-trapping gases; HFCs, for example, fell under MEP's umbrella. All of these authorities are retained under the new Ministry of Ecology and Environment.

In terms of fiscal policy, the Ministry of Finance (MOF; 财政部) is responsible for tax policy and has advocated for a carbon tax for a number of years. It also administers the resource taxes on oil, gas, and coal. MOF administers payments for China's feed-in tariff system for renewable energy as well, working together with the price bureau of the NDRC and the NEA to design and implement the FIT policy.

The Ministry of Information and Industry (MIIT; 工信部) formulates and implements most industrial policy. It issues climate-related policies aimed at improving industrial energy conservation, reducing industrial overcapacity, promoting industrial innovation and transformation, and implementing energy-efficiency standards in industry. The Ministry of Science and Technology (MOST; 科技部) promotes the development of science, technology, and innovation in China, along with the Chinese Academies of Science and Engineering (中科院) and the Ministry of Education (教育部). Although this book does not concentrate on the land-use GHG emission sources and sinks, the Ministry of Agriculture (农业部) has issued policies to control emissions from agricultural activities and to adapt to climate change. Relatedly, the State Forestry Administration (林业局) is responsible for enhancing forest coverage and wood stock.

The Ministry of Housing and Urban Development (MOHURD; 住建部), Ministry of Foreign Affairs (外交部), Ministry of Land and Resources (国土资源部), Ministry of Transport (交通部), and State-owned Assets Supervision and Administration Commission (SASAC, 国资委) are all also important players in the formulation and implementation of climate policy in China. As discussed earlier, the National Leading Group to Address Climate Change (国家应对气候变化领导小组) plays a key coordinating function.

In China, a *leading group* (领导小组) is a method or organization typically used to deal with broad and complicated issues that cut across the jurisdictions of multiple ministries. Two examples are the Leading Group for Food Security and the Leading Group for Health Care Reform. One of the most important roles of a leading group is to coordinate the different ministries' policies and actions, so the top leader of a leading group is usually the general secretary, premier, or vice premier, depending on the importance of the issue. Usually, leading groups have their own offices located within the ministry considered most important for that particular issue, such as the NDRC in the climate area or the National Health and Family Planning Commission for the health care reform area. The office is a key institution responsible for the regular operation of the leading group, including preparing its meetings.

The second period in target formation coincides with the timing of the eleventh five-year plan—specifically, 2006–2010. The Chinese government imposed an energy-intensity target for the first time during the eleventh FYP. This target required a 20 percent reduction in energy-onsumption levels in 2005 per unit of GDP by 2010. The energy-intensity target is a wonderful example of an indirect climate policy because although the main goal of the policy was to reduce waste in energy consumption, this target also had the benefit of reducing conventional air pollution and CO_2 emissions. According to one estimate, 4,273 Mt CO_2 emissions were likely avoided as a result of energy-efficiency policies in China during the eleventh FYP (Price et al. 2011).

Not only did the Chinese government establish an overall national energy-intensity target, but it also broke this target down for each major energy-consuming industry and province (see table 4.2). There is no precise mathematical process for breaking the national targets down by industry and province; it should be thought of as a negotiation process. For example, for a more developed province like Guangdong, the NDRC would likely set a stricter target than the average for the nation, but for a less-developed province, the target might be weaker. Provinces have the right to respond to the NDRC if they believe that the targets assigned to them are not appropriate, but in reality the provinces have little power to negotiate and ultimately must simply accept the targets. Furthermore, it is not the climate division in the NDRC that would assign the specific industry targets, but rather the relevant ministry or department. The transportation industry, for

Table 4.2
Energy-intensity targets by industry in China's eleventh five-year plan

	Unit	2000	2005	2010
Electricity generation				
Coal-fired	gce/kWh	392	370	355
Small generators	% (rated)	87	n/a	90–92
Wind turbines	% (rated)	70–80	n/a	80–85
Industry				
Raw steel	tce/t	0.906	0.760	0.730
Aluminum	tce/t	9.923	9.595	9.471
Cement	tce/t	0.181	0.159	0.148

Source: Adapted from Zhou, Levine, and Price 2010.

example, would receive its targets from the Ministry of Transport. The Ministry of Industry and Information Technology (MIIT) would supervise large manufacturing enterprises. In sum, the NDRC has the power to set targets for each province but does not have the authority to set targets for other ministries. As a result, the NDRC cannot ensure that the national target can be correctly implemented sector by sector, and this is a good example of China's "fragmented authority" discussed in chapter 3.

Provinces, autonomous regions, and municipalities directly under the central government subsequently established their own leading groups and working organs to address climate change, and some subprovincial or prefectural cities have also set up offices to address climate change (NDRC 2012).

In a move without parallel in the United States, the National Panel on Climate Change (国家气候变化专家委员会) was established in 2006 to provide expert input to the government. The panel was organized by the NDRC and the China Meteorological Administration (CMA). This committee of experts was comprised of university professors and experts from research institutes including the Chinese Academy of Sciences and Chinese Academy of Engineering. This committee was extended for two more terms, with a growing number of experts.

Relevant departments under the State Council also founded supportive research organizations, such as the National Center for Climate Strategy and International Cooperation, which was affiliated with the NDRC and worked closely with the Climate Change Department. The NCSC is analogous to the Energy Research Institute (ERI), which supports the National Energy Administration (NEA), which is also affiliated with the NDRC. The CCP and Congress also play roles in the climate policymaking process, but they do not engage in the day-to-day detailed work of climate policy. As of 2018, the NCSC also transferred its affiliation to MEE along with the climate change department of the NDRC.

During the eleventh FYP (2006–2010), China chose a strategy that combined its efforts to address climate change with implementation of a sustainable development strategy. This approach was integrated into the overall national economic and social development plan and regional plans. The concept of a *scientific approach to development* (科学发展观) was promoted by the central government, and the approach promoted a resource-saving and environmentally friendly society. This approach to development was consistent with the new National Climate Change Programme (中国应对气候变化国家方案) released in 2007, which defined the guiding principles, main fields, and key tasks concerning the work of addressing climate change. Notably, the target for controlling greenhouse gas emissions in the program included the same energy-intensity target as was in the eleventh FYP and the same target for achieving 10 percent of nonfossil energy in the primary energy supply by 2010. On the international side, President Hu Jintao attended the United Nations Summit on Climate Change in 2009 and delivered a speech titled "Join Hands to Address Climate Change," in which he set forth China's goals, positions, and opinions on addressing climate change (NDRC 2009).

During the third period in China's target development, the government set a carbon-intensity target for the first time. This period coincided with China's twelfth FYP, from 2011 to 2015, in which it was specified that CO_2 emissions/GDP would be reduced by 17 percent by 2015 as compared with 2010. This domestic target was consistent with the international target announced by the Chinese government in Copenhagen— specifically, a 40–45 percent reduction in carbon intensity by 2020 as compared with 2005. The twelfth FYP also included other targets: (1) energy-intensity reduction of 16 percent; (2) proportion nonfossil energy in the primary energy supply increased to 11.4 percent; (3) acreage of new forests to increase by 12.5 million ha; and (4) the forest coverage rate to be raised to 21.7 percent. A new National Plan to Address Climate Change (2014–2020; 国家应对气候变化规划) was compiled to guide the work for the coming ten years. Similarly, the "Work Plan for Controlling Greenhouse Gas Emissions during the 12th Five-Year Plan Period" ("十二五"控制温室气体排放工作方案) and "Comprehensive Work Plan for Energy Conservation and Emission Reduction during the 12th Five-Year Plan Period" ("十二五"节能减排综合性工作方案) documents were designed and implemented by the NDRC (see the appendix).

The second work plan was the successor of the plan in the eleventh FYP, but the first work plan focused on GHG emissions, assigning specific carbon-intensity-reduction targets to all provinces, autonomous regions, and municipalities directly under the central government. Based on the plan, the NDRC assessed and examined all of the different provinces to ensure that their efforts to reduce CO_2 emissions were on track to be successful.

The final period of China's target formation can be called the *emissions-peaking period*. This period began in 2014 when China first declared that its carbon emissions peak would occur "around 2030," with best efforts to peak early in the context of the Obama-Xi Joint Announcement. The NDRC coordinated the process of determining this target. Initially, the experts' group on climate change was asked to analyze possible targets and to present a range of views to Minister Xie Zhenhua (解振华). Input was sought from other ministries in China, and ultimately a recommendation was made through Vice Minister Zhang Gaoli (张高丽) to the State Council. Ultimately, the decision was made by Xi Jinping (习近平). The target-formation process in 2015 was extraordinarily short and circumvented a number of normal procedures to be completed in time for the 2014 Presidential Summit hosted by Xi Jinping in Beijing.

In 2015, China formalized the announcement when it submitted its INDC to the United Nations. The INDC also included other targets, such as a carbon-intensity reduction of 60–65 percent between 2005 and 2030, nonfossil energy supply of 20 percent by 2030, and a forest coverage target.

Domestically, continuing from to the twelfth FYP, China set a carbon-intensity target of 18 percent, an energy-intensity target of 15 percent for the 2016–2020 period, and a nonfossil target of 15 percent by 2020 in the thirteenth FYP. The central government also released the "Work Plan for Controlling Greenhouse Gas Emissions during the 13th FYP" Period ("十三五"控制温室气体排放工作方案) and "Comprehensive Work Plan for Energy Conservation and Emission Reduction during the 13th FYP

Period" ("十三五"节能减排综合工作方案) plans to achieve the target (see the appendix).

As this history reveals, there were two main forces behind the evolution of national target-formation policy in China to drive the policy through the evolution of the four stages—no compulsive target, energy-intensity target, carbon-intensity target, and emissions peaking. As in the United States, the first force came from the international side and the second force arose from the domestic side.

At the beginning, China insisted on its status as a developing country because China's per capita GDP was very low and its per capita emissions were very small. The Annex I countries of UNFCCC (industrialized countries) accounted for 78 percent of global CO_2 emissions from 1860 to 1990, and their population accounted for 22 percent of the world population in 1990. The non-Annex I (developing country) per capita CO_2 emissions were 0.48 t-C compared with Annex I emissions of 3.25 t-C. In 1990, China's per capita CO_2 emissions were just 0.55 t-C.

Like other members of the Group of 77 (G77), China emphasized the principle of common but differentiated responsibilities, which was embedded in the UNFCCC. This principle meant that industrialized countries should take the lead on emissions reductions because they were wealthier and had contributed the majority of GHG emissions in the atmosphere as of 1990. China also believed it was important to address climate change within the framework of sustainable development, which meant that China intended to place greater priority on development than on emissions control.

After several decades of high-speed growth, however, China became the top emitter and second-largest economy in the world. The international pressure became more and more intense as China's share of global GHG emissions grew dramatically. Even with its huge population, China's per capita emissions grew to be close to the per capita emissions of some East Asian and European industrialized countries, though not the United States.

For its own sake, China needed to exercise more leadership in the international community as its economic power strengthened. China always claimed to be "a country of responsibility" (NDRC 2007), and climate change offered a perfect stage for China to demonstrate that it could shoulder such responsibilities. In building good relationships with other countries, the climate change issue could be an area where China could be constructive. The decision to engage in negotiations with the United States in 2014 illustrated the Chinese government's intention to use climate policy as a diplomatic tool to promote cooperation and mutual trust with other major powers. Considering how many areas of conflict between the United States and China there were, the climate issue was seen as a good choice for cooperation between the two countries.

Domestically, during the 1990s, China was initially reluctant to take action to control its GHG emissions at home because of its pressing development needs. The first turning point occurred around 2005. China had entered into a new round of heavy industrialization driven by increased consumption of manufactured goods after 2000, which caused a large increase in energy consumption and related emissions. The energy intensity increased for the first time in the tenth FYP, although it had previously decreased ever since 1980. Although the Chinese government always said China needed to change its development model, which was characterized by "high inputs, high consumption, and high emissions (高投入、高消耗、高排放), it had made little progress since the ninth FYP. The central government needed a new weapon to push the transition of the development model, especially to change the behavior of local governments. As a result, the central government created energy-intensity and emission-reduction targets in the eleventh FYP. China established a mechanism to achieve these targets—specifically, the strict target responsibility evaluation and accountability system assigned to each level of government.

The second turning point was around 2010, when the Chinese government included a carbon-intensity target in the twelfth FYP. Of course, carbon intensity has a strong relationship with energy intensity, so it was somewhat redundant to do so, but including a carbon-intensity target demonstrated that consensus had been achieved internally that China was determined to tackle climate change directly. The mechanism created for implementing the carbon-intensity target was similar to that for energy intensity: a responsibility-evaluation system was established for local governments.

The third turning point was around 2015. At this time, the Chinese economy had entered into the so-called new normal phase, which meant that China could not maintain the super-high, double-digit pace of growth anymore. As a result, the economic growth rate could be expected to decrease to 6–7 percent. Although the energy-consumption and emission-growth rates had also decreased dramatically in recent years, which resulted in reductions in energy intensity and carbon intensity, air pollution had become a serious problem with adverse effects on ordinary persons' lives. Controlling air pollution became one of the government's top priorities, and it remains so today. Because air pollution comes mostly from energy consumption—especially from coal, which is closely related to greenhouse emissions—the climate issue continues to be connected with air pollution. The serious air pollution situation therefore helped push the government to adopt a stricter target in the climate domain.

In sum, the logic of national-target-formation policy in China is that the policies evolve as international and domestic forces change. China's logic has two characteristics. The first is continuity, which means that subsequent policies will not overthrow previous ones, new policies will inherit the important aspects of the old policies, and that the types of targets will be retained over time. The second characteristic is modification: new policies will not slavishly copy previous ones and will add something new based on environmental conditions. This means that increasingly stricter targets will be introduced step by step.

5 Target Implementation

Implementation of GHG emission-reduction targets is achieved through the use of many different types of policy tools at all levels of government. Typically, a full portfolio of approaches is required to effectively address all sources of emissions, across all sectors of the economy. Fundamentally, there are two types of climate policy instruments: direct and indirect. *Direct instruments* are those that explicitly target individual greenhouse gases for mitigation or adaptation. One example of a direct policy is a carbon tax, and another is the establishment of a target to reduce CO_2 emissions. A government would not impose a carbon tax or a target for any reason other than a desire to reduce CO_2 emissions (except, perhaps, a desire to increase government revenue). *Indirect instruments* are those that result in reductions of greenhouse gases but may not have been designed or implemented for that purpose. Indirect policies thus produce cobenefits (Smith and Haigler 2008) because they result in more than one kind of beneficial outcome. An example of an indirect policy is an energy-efficiency performance standard, which might be implemented to reduce energy consumption or to reduce oil or gas imports, but also results in reduced GHG emissions.

Policy instruments can also be classified by type: fiscal, regulatory/ administrative, market-based, informative, innovation, diplomatic, and other (see figure 5.1). *Fiscal instruments* use financial incentives or penalties to alter behavior—for example, a feed-in tariff or tax credit for renewable energy installations or deployments. *Regulatory/administrative policies,* though implemented very differently in the United States and China, can set conditions for emissions, such as a performance standard for automobiles that creates a maximum amount of CO_2 emissions allowed for each mile or kilometer driven. Another type of regulatory/administrative policy is to assign responsibility for reducing emissions to different regions of a

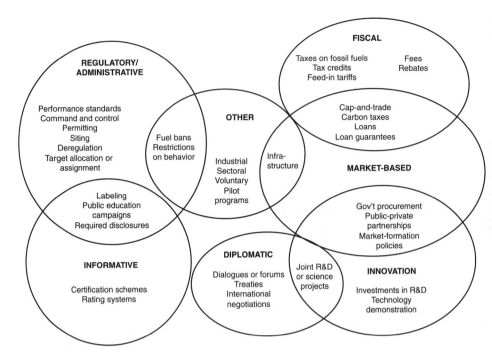

Figure 5.1
Types of climate-mitigation policies.
Source: Figure developed by K. S. Gallagher (2017).

country or different sectors. *Innovation* policies include investments into research, development, and demonstration of clean energy technologies, public procurement of advanced or low-carbon technologies, and initial subsidization of such technologies during the early commercialization process.

Market-based policies create new markets or set conditions for existing markets. A cap-and-trade program, for example, creates a market for permits to emit GHG emissions. *Informative policies* are aimed at public education or informing consumers about their purchases. The ENERGY STAR program in the United States, for example, requires manufacturers to put labels on appliances that explain how energy-efficient an appliance is compared with other models so that consumers can factor that information into their purchase decisions. *Diplomatic policies* are aimed at inducing cooperative behavior from other countries. The US-China Joint Statement of 2014 is an example of a diplomatic policy. There are many other types of policies that do not fit neatly into these categories, such as infrastructure policies, which may result in increased or decreased emissions. Of course, some policies may be classified in more than one of these categories. Industrial or sectoral policies can often cut across these categories, as shown in figure 5.1. Most of these policies can be targeted at particular sectors or regions and implemented at either the national or subnational level.

This chapter examines the target implementation approaches of the United States and China. The emphasis is primarily on emission-reduction (mitigation) policies, not on policies for adapting to climate change. This allows us to examine how the targets adopted as discussed in chapter 4 are implemented in each country.

Some policy instruments, as implemented, are very similar in the United States and China. The US ENERGY STAR label, for example, has a similar equivalent in China. Other types of policies are promulgated and implemented very differently due to many institutional, structural, political, and procedural differences. China's target-oriented responsibility and assessment system, for example, is unique to China and impossible to imagine in the US context.

US Implementation

Pre-2009
At the national level, there were no serious national efforts to directly reduce GHG emissions by any administration before that of President Obama, but a number of policies had indirect and substantial effects in reducing GHG emissions below what they otherwise would have been. The most important indirect policies were those intended to spur the deployment of renewables, to catalyze energy-efficiency actions, and to incentivize energy technology innovation.

Renewables Originally enacted in 1992, the corporate production tax credit (PTC) for renewable energy was historically the main policy tool employed to incentivize wind and solar energy in the United States. The renewables PTC is a perfect example of an indirect policy because the motivation and political support for it has varied over the years, sometimes being primarily motivated by energy-security concerns, sometimes by a desire to create "green jobs," and otherwise intended as a means to

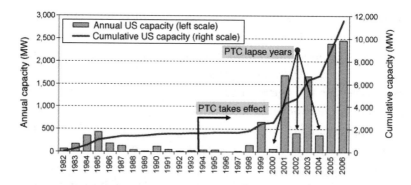

Figure 5.2
Boom-and-bust cycles for wind energy in the United States.
Note: The first PTC lapse actually lasted only until mid-December 1999 (not 2000), but the late-1999 renewal impacted development activity in 2000.
Source: Wiser, Bolinger, and Barbose 2007.

reduce conventional air pollution. The US Congress has allowed the PTC to expire many times, even though it has always eventually been reauthorized. Between 1998 and 2008, for example, the PTC for wind lapsed six times, as depicted in figure 5.2, which both challenged and undermined the renewable energy industry because the unpredictability led to repeated boom-and-bust cycles (Barradale 2010).

Although the precise amount of tax credit for different types of renewable energy has changed over the years, the PTC has always been administered as an "inflation-adjusted per-kilowatt-hour (kWh) tax credit for electricity generated by qualified energy resources and sold by the taxpayer to an unrelated person during the taxable year" (DOE 2017); as of this writing, it applies to the first ten years of a facility's operation. As of the 2016 Consolidated Appropriations Act, the tax credit was USD 0.023/kWh for wind energy, which is phased down over time. During the Obama administration, an investment tax credit (ITC) was also provided in the American Reinvestment and Recovery Act (ARRA), as discussed later in this chapter.

Congress has never passed a national renewable portfolio standard (RPS) due a lack of political consensus, but at the state level, RPSs, use of public benefits funds, and electricity market deregulation and restructuring all provided additional incentives for renewable energy before the Obama administration. At the state level, the public supported these policies because clean energy was seen as good for the environment and a creator of new jobs. Even in politically conservative states like Texas, strong RPSs were created (Ansolabehere and Konisky 2012). In Texas, the legislature established a renewable generation requirement starting in 2000 that mandated five thousand megawatts (MW) of new renewables by 2015 and 10,000 MW of renewable energy capacity by 2025. According to the Electric Reliability Council of Texas (ERCOT), Texas surpassed the target it set in 2000 for 2025 much more quickly than anticipated (in 2009); as of 2013, it had 13,359 MW of additional renewable energy capacity (wind accounting for the majority) relative to 1999 (DOE DSIRE 2016).

Efficiency   The American Council for an Energy-Efficient Economy estimated that US energy-efficiency policies avoided the need to build the equivalent of 313 large (500 MW) power plants between 1990 and 2015, reducing annual CO_2 emissions from the power sector alone by 490 million tons in 2015 (Molina, Kiker, and Nowak 2016). The federal policies that were used to achieve these changes included appliance- and equipment-efficiency standards, EPA's ENERGY STAR labeling program, and federal energy-efficiency procurement programs (government purchases of energy-efficient products for use in government buildings). The Department of Energy and EPA have authorities under laws previously passed by Congress to implement and update these types of policies.

Although they were not updated for decades, the original fuel-economy standards for automobiles enacted in 1975 also avoided substantial future oil consumption and greenhouse gas emissions. A 1998 study of the corporate average fuel economy (CAFE) program calculated that if the 1975 standards had not been implemented, vehicles in 1998 would be consuming fifty-five billion more gallons of fuel per year at a cost of USD 70 billion (in 1995 dollars; Greene 1998).

Congress passed the Energy Independence and Security Act (EISA) in 2007, which improved energy efficiency for lightbulbs by establishing a process for setting progressively higher efficiency performance standards through 2020 and led to the phase-out of old-fashioned incandescent light bulbs (EPA 2017d). EISA also authorized updated efficiency standards for residential boilers, fans, air conditioners, heat pumps, freezers, and so forth, and created a number of incentives for energy efficiency.

Energy Innovation   Although political consensus regarding explicit climate change policy has never existed in the United States, bipartisan support for investing in energy technology innovation has historically existed. Of course, the total amounts invested and energy technology preferences have varied considerably over the years, as evident in figure 5.3. Public investment in energy research, development, and demonstration (RD&D) in the United States peaked in 1978 in the wake of the oil crisis at nearly USD 9 billion in 2015 dollars and with strong support from the Carter administration for a broad portfolio of approaches ranging from nuclear fission to renewables and even coal-to-liquids. After President Reagan was elected, US government investment in energy RD&D dropped precipitously to less than half of the 1978 peak and support for renewables, efficiency, and nuclear fission technologies flattened out, while investments for "clean coal" were expanded during the 1980s and early 1990s. Investments remained essentially flat during the Clinton administration, although the nuclear fission investments almost disappeared at the end of the 1990s. With the George W. Bush administration, new initiatives in hydrogen energy and renewed interest in clean coal and nuclear energy led to a modest but temporary bump in energy RD&D investments.

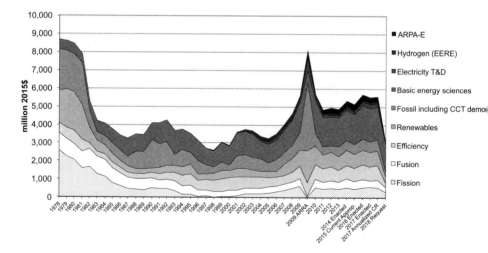

Figure 5.3
US DOE energy RD&D FY 1978–FY18 request.
Source: Gallagher and Anadon 2017.

Obama Administration

Explicit climate policy in the United States commenced under the Obama administration, spurred by a dozen progressive US states. During the George W. Bush administration, Massachusetts filed a lawsuit on behalf of eleven other states, three cities, one territory, and a number of public-interest NGOs against the federal EPA for failure to regulate carbon dioxide as a pollutant under the Clean Air Act, arguing that CO_2 created a threat to public health and welfare. The US Supreme Court ultimately upheld this lawsuit in 2007 in a 5 to 4 vote (Greenhouse 2007). The states suing EPA argued they had standing to do so because the citizens of their states were being endangered by the lack of federal policy on climate change. Massachusetts argued, for example, that it was losing coastal property due to rising sea levels. According to section 202(a)(1) of the US Clean Air Act originally passed by Congress in 1970 and revised in 1977 and 2000, the administrator of the EPA is required to set emission standards for "any air pollutant" that causes or contributes to "air pollution which may reasonably be anticipated to endanger public health or welfare" (US Code 2017b). The administrator under Republican George W. Bush had previously determined that GHGs should not be considered air pollutants and therefore that the EPA was not obliged to regulate them. The Supreme Court majority disagreed with this determination and therefore obligated the EPA to regulate GHGs (Greenhouse 2007).

When the Obama administration came into power, the new EPA administrator was obliged by the Supreme Court ruling to initiate regulatory processes to regulate CO_2 from major sources. The first major EPA rule-making process on climate change created CO_2 performance standards for motor vehicles. The first phase reduced CO_2 emissions from new cars and light trucks from 2012 to 2016, and the second phase was scheduled for new cars sold between 2017 and 2025. According to the EPA, these regulations were estimated to result in an average industry-fleet-wide level of 163 grams/mile of CO_2 emissions in model year 2025, which is equivalent to an energy-efficiency standard of 54.5 miles per gallon (mpg) and the avoidance of six billion metric tons of GHG over the lifetimes of the vehicles sold in model years 2012 to 2025 (EPA 2016a).

The new CO_2 performance standards for motor vehicles were the main accomplishment on climate change for the Obama administration's first term. One indirect regulation may also prove to have a lasting impact: the Mercury and Air Toxics Standards (MATS) rule. MATS requires all oil and coal-fired power plants, new and old, to limit emissions from mercury, arsenic, and other toxic metals (EPA 2017e). Although various control technologies exist to reduce these types of emissions, they add to the cost of operating a coal-fired power plant and thus create an additional disincentive for construction of new coal-fired power plants. As previously discussed in chapter 4, major climate legislation was attempted in the US Congress during President Obama's first term, but it did not pass the Senate due to significant political opposition.

After President Obama was reelected in 2012, he reportedly decided to make climate change policy a high priority for his second term (Broder 2013). Congressional legislation had failed, so it was apparent to his administration that they would need to utilize existing authorities under existing laws if they wanted to reduce GHG emissions in the United States. In June 2013, President Obama announced a new Climate Action Plan (CAP), arguing that "we have a moral obligation to leave our children a planet that's not polluted or damaged" (White House 2013). The CAP had three components: "cut carbon pollution" (mitigation), "prepare the United States for the impacts of climate change" (adaptation), and "lead international efforts to address global climate change" (diplomacy).

Because of the lack of political support in Congress for comprehensive climate legislation, discussed in chapter 4, the Obama administration had no choice but to primarily rely upon regulatory, informative, diplomatic, and innovation policy tools to implement the US target of a 17 percent reduction below 2005 levels by 2020. Fiscal and market-based policy instruments require approval by Congress, which holds the power of the purse under Article 1, Section 8 of the US Constitution. This means that only Congress has the authority to tax the public and spend public money.[1] A US president therefore can not implement a federal carbon tax or cap-and-trade program without authorizing legislation from the Congress, but such policies can be implemented at the state level if state legislators pass their own legislation. Under existing federal laws, including the Clean Air Act, substantial authority already existed for President Obama to begin to reduce emissions using other policies, and he was also able to compromise with Congress on some other important indirect climate policies.

Regulatory/Administrative Policies   With the announcement of the CAP, the president directed the administrator of the EPA to use her authority under the Clean Air Act to regulate GHG emissions from power plants. The EPA administrator, Gina McCarthy, immediately initiated a rule-making process for establishing carbon performance standards for power plants in what became known as the Clean Power Plan, which would have reduced CO_2 emissions from power plants 32 percent below 2005 levels by 2030. This rule-making process took years; the final rule was issued in 2015, but it was immediately challenged in the courts by twenty-seven opposing states. Approximately fifty additional lawsuits have been filed against the Clean Power Plan (E&E News Reporter 2017). The Supreme Court took the extraordinarily unusual step of granting a stay, which meant that EPA could not proceed with implementing the rule until the lawsuits were settled (Stohr and Dlouhy 2016). The lawsuits surrounding the Clean Power Plan will proceed through lower courts and eventually will be settled by the Supreme Court. Given that President Trump was able to nominate the latest Supreme Court justice, a conservative judge, the composition of the Supreme Court is now different from when it ruled that GHGs must be considered a pollutant under the Clean Air Act. The new composition of the Supreme Court means that a majority of justices most likely will not uphold the Clean Power Plan.

Other major regulations planned under Obama's CAP included permitting renewable energy on federally owned land, developing heavy-duty vehicle CO_2-performance standards, tightening efficiency standards on appliances, and reducing HFCs under the Significant New Alternatives Policy (SNAP) program at EPA.

President Trump's EPA administrator, Scott Pruitt, has systematically tried to halt or reverse many of the regulatory steps taken under President Obama. In March 2017, the EPA announced it intended to withdraw its final determination on stricter fuel-economy standards for new cars sold between 2022 and 2025 (Eilperin and Dennis 2017). Similarly, Administrator Pruitt tried to halt implementation of the Obama-era regulation requiring oil and gas companies to reduce leaks of methane. However, new lawsuits have been filed by states and NGOs to counter the regulatory rollback under the Trump administration. Administrator Pruitt suffered his first defeat when the US Federal Court of Appeals ruled in July 2017 that the EPA could not halt implementation of the methane rule (US Court of Appeals 2017).

The regulatory approach in the United States is thus fraught with legal risk because if a new rule issued by EPA or another federal agency does not have broad support from the public and regulated industries, it can be tied up in the court system for years. The use of the court system to influence the implementation of regulatory actions in the United States cuts both ways: although progressive climate regulations can be slowed down or thrown out in the judicial process, it is also true that attempts to unravel existing regulations can be slowed down or halted.

Innovation Policies   The portfolio of government investments in energy RD&D was drastically altered during the Obama administration under the direction of Secretaries of Energy Steve Chu and Ernest Moniz and Director of the White House Office of Science and Technology Policy John Holdren, with support from the Office of Management and Budget. The Obama administration consistently requested larger energy innovation budgets than were ultimately appropriated by Congress. On a percentage basis and adjusted for inflation, the new Advanced Research Projects Agency-Energy (ARPA-E) grew the most (2,896 percent) between fiscal years 2009 and 2016, followed by energy efficiency RD&D (20 percent) and basic energy sciences (8 percent). Hydrogen energy suffered a cut of 45 percent, and fossil energy investments were also curtailed by 35 percent. Renewables and nuclear investments held steady during this period. Overall, energy RD&D plus basic energy sciences grew just 1 percent during the Obama administration, not counting the major infusion of funding from the American Reinvestment and Recovery Act. The ARRA investments in energy innovation were relatively enormous, as shown in figure 5.2, totaling more than USD 8 billion and approaching the historic peak from 1978.

President Obama also utilized procurement policies for federal buildings to incentivize cleaner and more efficient energy technologies through executive orders that set targets for the use of different types of energy or equipment. The Department of Defense also began procuring biofuels and renewable energy for military equipment and installations around the world.

Fiscal Policies   With support from Congress, several important fiscal policies had indirect influence on carbon emissions during the Obama administration. The first fiscal policy, tax credits for renewable energy, was most recently extended in the Congressional Appropriations Act of 2016. Production tax credits were extended for wind, geothermal, biomass, hydro, municipal solid waste, landfill gas, tidal, and wave energy. The solar industry now relies upon investment tax credits; as of the end of the Obama administration, it provided a 30 percent tax credit for solar systems on residential and commercial properties. The company that invests USD 100 to install, develop, or finance a new system would thus receive USD 30 in a tax credit. The Emergency Economic Stabilization Act of 2008 (PL 110–343) included an eight-year extension of the commercial and residential solar ITC, so an ITC was in place for the entire Obama administration, greatly contributing to the expansion of this industry in the United States (SEIA 2017). Tax credits were also provided for electric vehicles throughout the Obama administration and were administered as personal income tax credits for individuals purchasing EVs, which means that they are received when such a person files his or her income taxes.

The second influential fiscal policy in support of clean energy was the provision of loan guarantees. Although they were authorized in the Energy Policy Act of 2005, such guarantees were greatly expanded in the American Reinvestment and Recovery Act of 2009, which provided nearly USD 4 billion to the Title 1705 loan guarantee program administered by the US DOE (Gallagher and Anadon 2017). Because these loan guarantees were supported as a tool for economic recovery, construction on renewable energy or electricity transmission and distribution projects had to commence no later than 2011 (DOE n.d.-b). An earlier loan guarantee program for the automobile industry was also devised to support economic recovery, and this was called the Advanced Technology Vehicle Manufacturing; loan guarantees were extended to manufacturers who intended to expand production of more fuel-efficient vehicles. As of September 2016, USD 30 billion in loans and guarantees had been disbursed, according to the DOE (DOE n.d.-b).

Diplomatic Policy   The third component of President Obama's 2013 climate action plan was devoted to international policy, and it called specifically for "leading international efforts" to address global climate change. Many accounts of President Obama's personal experience at the 2009 Copenhagen conference of the UNFCCC indicate that he was both surprised and troubled by the chaotic state of the international negotiations.

President Obama subsequently provided strong encouragement to Secretary of State John Kerry and Special Envoy for Climate Change Todd Stern to take a more active role in the negotiations. It was not easy for them to do so given that the United States had little credibility on this issue internationally because it had negotiated but then failed to ratify the Kyoto Protocol. Still, once the other components of President Obama's Climate Action Plan began to be implemented, Stern stood on firmer ground and could negotiate with greater moral authority. One approach strongly advocated by Stern was the idea of *pledge and review* (as discussed in chapter 4). This bottom-up approach to establishing emission-reduction targets removed a major political impediment for the United States—specifically, that all countries would specify targets suitable for their own circumstances and thereby make the ultimate agreement a universal one. The main reason that the US Senate had refused to ratify the Kyoto Protocol was because major developing countries—in particular, China—did not have obligations under that agreement.

As discussed in chapter 1, Secretary Kerry took the initiative to propose the idea of a joint announcement to the Chinese government, which was subsequently negotiated by John Podesta from the White House and Todd Stern from the US State Department. Many believe that the 2014 US-China Joint Announcement (and 2015 joint statement) made the Paris Agreement possible (see, for example, UN Climate Change Newsroom 2014).

In 2014, President Obama personally appeared at the UN Climate Summit convened by UN Secretary General Ban Ki Moon in New York to make a speech. In that speech, President Obama recognized the special concerns of many developing countries about how they will be able to adapt to climate change and announced a new public-private partnership to provide climate services (similar to weather services) to developing countries. This partnership, Climate Services for Resilient Development, was launched in 2015 together with the UK government, Esri, Google, American Red Cross, Skoll Global Threats Fund, Asian Development Bank, and the Inter-American Development Bank. Led by the US Agency for International Development (USAID), most of the major science and technology agencies of the US government, including the National Aeronautics and Space Administration (NASA), the National Oceanic and Atmospheric Administration (NOAA), and the US Geological Survey (USGS), agreed to contribute climate data and tools to meet the information needs of focus countries (Office of the Press Secretary of The White House, 2015).

Trump Administration

We finished this book during the first year of the Trump administration, so it was too early to fully assess its impact. President Trump honored a campaign promise and announced he intended to withdraw the United States from the Paris Agreement on Climate Change, even though many Republican senators and business leaders urged him not to do so. As discussed previously, this symbolic step was preceded and followed by a systematic effort to undo the regulatory actions initiated under President Obama's Climate Action Plan, including the Clean Power Plan, the methane rule to limit leaks of CH_4 from oil and gas infrastructure, and the newest light-duty vehicle fuel-economy standards.

Environmental NGOs and progressive US states sued the federal EPA, DOE, and DOI under President Trump to try to halt the regulatory rollback (HLS 2018). These lawsuits may ultimately prove successful given that the Supreme Court already ruled that CO_2 and other GHGs must be regulated under the Clean Air Act, but it could take years for these lawsuits to be resolved. The Supreme Court is now less likely to support environmental cases, however, because President Trump was able to nominate the tie-breaking judge on the Supreme Court after his election.[2] Previously, there were four liberal and four conservative judges, with one more independent judge who swung back and forth depending on the issue. With President Trump's new appointment, confirmed by the Senate, the conservative side now has a majority, assuming the new justice, Neil M. Gorsuch, sides with the conservatives on environmental cases. Although Supreme Court justices are supposed to be apolitical, the fact of the matter is that because they are nominated by the president and must be confirmed by Congress, their political views are usually well known.

Many regulations remain unchallenged by the Trump administration as of this writing, including most of the energy-efficiency standards issued by the US Department of Energy. The production and investment tax credits for most renewable energy technologies passed by Congress in the Consolidated Appropriations Act of 2015 will remain in place for the duration of President Trump's first term (DOE n.d.-a, n.d.-d). Finally, because Congress did not support the draconian cuts to the US energy-innovation budget proposed by President Trump in his first budget request to Congress, it is likely that the US government energy innovation investments will experience only modest cuts during the Trump administration (DOE n.d.-e).

Overall Mitigation Impact of US Policies

Putting all of the mitigation policies discussed so far together, plus others listed in table 2.8, the Obama administration estimated that a 17 percent reduction below 2005 levels was likely to be achieved prior to 2020 (Obama's Copenhagen target) and that the Paris target of 26–28 percent below 2005 levels could be achieved by 2025 with an optimistic land-use sink and additional new policies (see figure 5.4) (US Department of State 2016). The Obama administration submitted this figure to the UNFCCC to demonstrate business-as-usual cases (labeled "Current Measures") and the emissions impact of its new Climate Action Plan policies. The automobile standards and other energy-efficiency standards from the first term clearly are shown to cause a reduction in GHG emissions. Assuming that some of the regulatory and fiscal policies survive the Trump era, US emissions will likely to continue to fall, but not as rapidly as they would have if the regulations had not been tied up in the courts.

This Obama-era projection assumed full implementation of all of the policies implemented under the Climate Action Plan, including the Clean

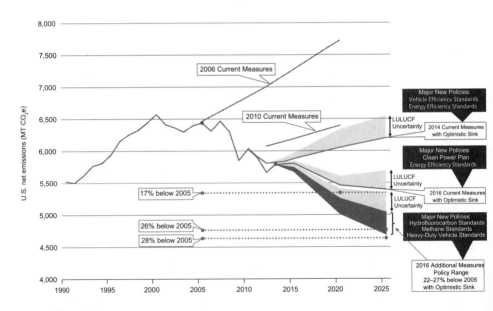

Figure 5.4

US emissions projections: 2016 measures compared with potential reductions from additional measures consistent with the climate action plan.

Power Plan, the methane rule, reduced HFCs, and many energy-efficiency standards. No additional policies were anticipated at that time and included in the analysis.

Despite President Trump's hostility toward climate regulations, a number of the existing policies are unlikely to be reversed because they have enjoyed bipartisan support over the past decades. According to the 2016 US Biennial Report submitted to the UNFCCC, eighty individual policies implemented in the United States are providing climate change benefits (US Department of State 2016). Some of those policies date back to as early as the 1970s, and others were initiated during Republican administrations.

During the Reagan administration, there were incentives for energy-efficient homes (1985), the conservation reserve program that encouraged farmers to conserve erodible cropland (1985), and the initiation of appliance- and equipment-efficiency standards (1987). Under George H. W. Bush, SNAP was initiated in 1990, and that policy is responsible for shifting away from ozone-depleting chemicals, some of which are also greenhouse gases. SNAP is expected to deliver more than three hundred million metric tons of CO_2eq in emission reductions in 2020 alone (US Department of State 2016). During the George W. Bush administration, the Carbon Sequestration Leadership Program was initiated to demonstrate CCS technologies, the SmartWay Transport Partnership was implemented to reduce emissions from the movement of goods (2004), a biofuel feedstock partnership supported the supply of biomass, and an HFC reduction program was implemented in 2006. Admittedly, few of these policies have proven to be major contributors to reducing GHG emissions (with the notable exception of SNAP), but the history shows that very few previous policies are reversed or canceled by subsequent administrations.

State-Level Policy

In addition to the federal policies discussed so far, US states and municipalities have enormous influence over US greenhouse gas emissions through their own policies. States and municipalities are allowed to pass more aggressive legislation than the federal government's so long as they are not inconsistent with federal policies. Just as climate policies can be categorized as direct or indirect at the federal level, the same is true at the subnational level.

At the state level, Texas is the overall largest CO_2 emitter, followed by California, Pennsylvania, Florida, and Illinois (in that order). Small states Vermont, Rhode Island, and Delaware are the smallest emitters, although it should be pointed out that all three are part of the Regional Greenhouse Gas Initiative (RGGI), discussed ahead, and therefore have made significant efforts to reduce their emissions. Seven states have reduced their emissions more than 20 percent between 2000 and 2015: District of Colombia, Indiana, Maine, Massachusetts, Nevada, Tennessee, and New York. Delaware, Ohio, North Carolina, and West Virginia are close behind (EIA 2018b).

Historically, some of these subnational governments have been far more progressive than the federal government, experimenting with climate policies many years before they are ever contemplated nationally. Over time, as more and more cities and states have adopted climate policies, a larger geographic fraction of the United States has become governed by climate policy. Every single US state except Alaska had at least thirty clean-energy policies as of 2017, and some states many more (California had 272) (DOE DSIRE 2016). As of 2017, thirty US states had renewable portfolio standards in place. Six months after President Trump's inauguration, California's legislature passed a new law extending California's cap-and-trade program through 2030. This program imposes a declining state-wide cap on GHG emissions (Nagourney 2017).

Politically conservative states typically utilize indirect climate policies, including energy-efficiency programs like Mississippi's commercial energy-efficiency program, Texas's renewable energy requirement (a renewable energy portfolio standard), or Alaska's building energy codes. Politically progressive states are increasingly adopting explicit climate policies, such as California's cap-and-trade program for greenhouse gases or the City of Boston's plan to reduce vehicle miles traveled 5 percent below 2010 levels by 2020 through expansion of public transit and bicycle lanes as a measure to reduce GHG emissions.

What is not possible politically at the national level is often possible at the state level because the attitudes toward environmental protection (and climate change specifically) vary dramatically across states; therefore, the politics within each state are completely different from the politics of the nation as a whole. Nationally, the conservative states have greater influence because of the composition of the US Senate, which has two senators from each state. By contrast, the House of Representatives is comprised of Congressmen elected based on population; because a majority of the US population supports climate change action, the House has been more progressive on climate change issues than the Senate. You will recall from chapter 4, for example, that the Waxman-Markey climate change bill was able to pass the House of Representatives, but its equivalent was not able to pass the US Senate.

Even conservative states and their leaders have supported renewable energy so long as climate change was not the rationale. As governor, George W. Bush signed into law the Texas Public Utility Regulatory Act of 1999, which led to a huge expansion of wind energy in Texas. In his comprehensive book on state-level climate policy, Barry Rabe (2004) argues that policy entrepreneurs at the state level successfully linked climate policy to long-term economic development and jobs.

Still, some states simply seem to have greater political will for progressive climate policy than others, especially in the more liberal parts of the country: New England and California. Both of these regions have implemented cap-and-trade programs. In New England, the Regional Greenhouse Gas Initiative was the first such program for carbon dioxide in the United States, and it is a cooperative program among nine states to cap and reduce CO_2 emissions from the power sector. Between 2005 and 2016, power-sector emissions were reduced by 45 percent (1.7 million tons between 2008 and 2016) even while the regional economy grew 8 percent. The auction revenue was reinvested in each state into a variety of energy-efficiency, clean-energy, GHG-abatement, and direct bill-assistance programs (RGGI 2016). In 2017, the RGGI states announced a further 30 percent reduction of their cap between 2020 and 2030.

The California cap-and-trade program is even more comprehensive (covering all GHGs, major industrial emitters, the power sector, and fuels distributors) and started in 2012. In 2014, California linked its cap-and-trade program to that of Quebec, Canada. The California Global Warming Solutions Act of 2006 required the California Air Resources Board to approve a statewide greenhouse gas emissions limit equivalent to the statewide greenhouse gas emissions level in 1990 to be achieved by 2020 and to ensure that statewide greenhouse gas emissions are reduced to at least 40 percent below the 1990 level by 2030. The act authorizes the state board to include the use of market-based compliance mechanisms (Garcia

2017). As of 2014, California's emissions were 1.6 percent below 1990 levels (EIA 2018b).

Market Forces and Structural Economic Change

Market forces and structural economic change have contributed to reducing US GHG emissions, but there is considerable debate about their relative influence. The shift in electricity from coal to natural gas, and then again to renewable energy sources, caused US GHG emissions to begin to fall after 2005. Shifts from nuclear to gas offset some of these gains because nuclear is almost carbon-free. According to one study, changes in the fuel mix accounted for 4.4 percent of the 11 percent reduction in emissions between 2007 and 2013. Changes in economic structure were even more important, accounting for 6.1 percent of the 11 percent reduction during this period (Feng et al. 2015).

The government did not explicitly direct this shift from coal to gas through policy; rather, the shift occurred when independent power producers chose to close coal-fired power plants and shift to natural gas as a fuel source because the price of natural gas fell dramatically after US domestic shale gas began to be produced. CO_2 emissions from the combustion of natural gas are much smaller than from combustion of coal, but when leakage of natural gas from the production and distribution infrastructure is taken into account, the climate benefits are smaller—or even net positive, depending on the amount of leakage, given the global warming properties of methane. It is thus important to limit leakage of methane from oil and gas production and distribution in order to maximize climate benefits from fuel switching in the power sector (Alvarez et al. 2012).

These market forces were partly unleashed by US government investments in energy innovation and by public-private partnerships: the early shale-fracturing and directional-drilling technologies developed by the Energy Research and Development Administration (later the DOE), the Bureau of Mines, and the Morgantown Energy Research Center; the Eastern Gas Shales Project (a public-private shale-drilling demonstration program in the 1970s); public subsidization of demonstration projects, including the first successful multifracture horizontal-drilling experiment in West Virginia in 1986; and Mitchell Energy's first horizontal well in the Texas Barnett shale in 1991, among other collaborations (Trembath et al. 2012).

Lessons Learned about US Climate Policy

Three important lessons can be drawn from this examination of how climate policies have been implemented in the United States. First, no president has the constitutional power to ensure that he or she can honor a commitment made in international negotiations if the commitment extends beyond his or her administration and if the US Senate has not ratified the treaty or the Congress has not passed implementing domestic legislation. The regulatory approach used by President Obama can be effective but is proving easily contested in the courts.

Second, deliberate and explicit domestic climate change policy did not begin until the Obama administration, but many indirect climate policies provided climate benefits long before then, and these policies were supported by both political parties at both the federal and state levels.

Third, all of the different types of climate policies utilized by the US federal government and by the states combine to create a comprehensive policy regime, but it is fragmented geographically, unstable, and unpredictable for firms and citizens alike.

Chinese Implementation

Like in the United States, the Chinese government did not release explicit climate policies until the thirteenth five-year plan (2015–2020), but many dozens of indirect climate policies were issued during the 1990s and 2000s (Li et al. 2016), as discussed in chapter 2. The main types of policy instruments used by China's policymakers are regulatory/administrative, fiscal, market-based, and industrial (which is included in the "other" category in figure 5.1). Ahead, we endeavor to focus on the policies that have been most influential in reducing GHG emissions, according to a study that identified the top twenty most important policies to reduce CO_2 in China (Gallagher, Zhang, and Orvis 2018).

For the sake of comparison, we will use these aforementioned policy types to discuss Chinese implementation during the rest of this chapter, but a different typology might emphasize four main Chinese approaches to climate policy: target allocation, a long-term campaign to "upgrade" the energy mix, experimentation, and market-based.

The target-allocation system is the starting point for climate policy in China. The central government decides on national targets in the context of its five-year plans, and then it breaks the targets down among provinces. The provinces, in turn, break their targets down among counties and municipalities within each province. Although the targets for lower levels of government can be negotiated to some extent, this target-allocation system is fundamentally a top-down process, discussed further ahead in the context of regulatory/administrative approaches (Li, Li, and Qi 2011). There is no exact science for the decisions, and they may vary over time given particular circumstances. If the air pollution is very serious or it is very urgent to meet the target of energy intensity or carbon intensity, there is little room for lower governments to negotiate with higher levels, in which case the higher-level government will have absolute power to decide the target and the lower-level government must passively adopt the target assigned to it.

The central government's campaign to diversify the sources of energy supply is a long-standing one, aiming to reduce use of coal and increase use of renewables, natural gas, and nuclear power. Many specific policies were implemented in the context of this "upgrading" campaign, including establishing a cap on overall coal consumption, implementing feed-in tariffs for renewable energy, establishing specific targets for all types of energy, upgrading the grid so it could accept more renewable energy, creating an overall target for the percentage of nonfossil energy supply, reducing subsidies for fossil fuels, and reforming the power sector.

Increasingly, China's government is now turning to market-based approaches to climate policy. After implementing seven pilot cap-and-trade programs at the provincial and city levels in China, China announced in 2015 that it would scale up to a national emissions-trading system (ETS) by the end of 2017. This program is initially focused on the power sector and includes 1,700 enterprises and covers 3.3 billion tons of CO_2 emissions. In addition, China implemented an environmental tax that does not include CO_2 but may indirectly cause reductions of CO_2 and other GHGs.

The central government increasingly recognizes advantages to market-based approaches because of the challenges it has experienced implementing regulatory and fiscal policies in a country as geographically large and populous as China. The lack of direct enforcement tools is a key challenge for the central government. Its main tool for compliance is to not promote party officials if they fail to implement the policies. It is not common to use the court system in China to address noncompliance with environmental policies, nor does China's central government have enforcement institutions at the local level. As discussed in chapter 3, one of the most important characteristics of the central-local government relationship in China is that the central government is responsible for the policy decisions and local governments are responsible for the implementation.

As we can see from the cap-and-trade example, the final characteristic of China's approach to climate policy is experimentation. The central government will almost always experiment with a policy with a limited set of participants first, then scale it up if the experiment appears to work. Experimentation through pilot projects or regions is a typical approach. Pilot projects not only allow the government to try out different policy approaches but also are one of the few ways that the central government ministries can play a direct role in the local government given the institutional arrangement between the central government and local governments.

Indirect Policies before the Twelfth Five-Year Plan

As discussed in chapter 2, many of China's policies originate in the context of five-year plans and then are updated with each successive one. As early as 1986, the central government was focused on improving energy efficiency, by issuing design performance standards for buildings that year. In this section, we examine China's regulatory/administrative, fiscal, industrial, and innovation policies.

Regulatory/Administrative As mentioned previously, the target-establishment and -allocation administrative system is the starting point for most climate policy in China. Overall goals are set for the country, and then specific national targets are chosen that will lead to the achievement of those goals. Because the NDRC has authority over most energy and climate change issues, it is responsible for determining the specific targets for energy intensity, nonfossil energy share, and specific capacities for different types of energy. The NDRC divides the national target into subnational targets, assigning one to each province and taking into account the economic development status of each region. The NDRC typically asks provinces to propose their own targets, but because none wish to put too much pressure on themselves, they will submit ones that are too modest. The NDRC will take their views into account and hand down the final targets. This back-and-forth can be a protracted process of negotiation, but it can be accelerated if the leadership feels urgency. The provinces then must go through the same process of breaking the provincial target down for all of the cities and counties within the province, assigning specific targets to each one. There is huge variation in how each province decides to allocate its target. The target-allocation enforcement is primarily achieved through the cadre-evaluation system.

Dozens of targets are articulated in each five-year plan that affect energy consumption and GHG emissions, including annual economic growth targets, energy-intensity targets, targets for the proportion of nonfossil fuels in the primary energy supply, and specific targets for natural gas, renewable energy, and nuclear energy power capacities, to provide a few examples. Establishment of targets is crucial because it clarifies the objectives for policymakers and industry leaders at every level of governance in China. But the existence of a target is insufficient to achieve the objective; implementing policies are required.

For achievement of the energy-intensity targets, the regulatory/administrative tools used by the Chinese government include fuel-efficiency performance standards for buildings, appliances, industrial equipment, motor vehicles, and manufacturing. The government also implemented the Top-1,000 Energy-Consuming Enterprises Program, which was later expanded to the Top-10,000 Energy-Consuming Enterprises Program. In the context of this policy, energy conservation targets were set for large industrial energy consumers. In exchange for meeting these targets, firms were provided with subsidies to purchase more energy-efficient equipment. During the tenth FYP (2001–2005), economic growth was proceeding at a very rapid pace, and the energy intensity of the economy actually began to increase; therefore, in the eleventh FYP (2006–2010), energy intensity became a target for the first time. The target was a 20 percent reduction in energy intensity, but only 19.1 percent was achieved. Premier Wen Jiabao (温家宝) led an intensive campaign to reduce energy intensity. Local officials were under such pressure to achieve the target that restrictions were imposed on industrial energy consumption; in one extreme case, local government officials arbitrarily cut power to a hospital to meet the target (Yuan and Feng 2011). Small and inefficient energy-intensive factors and coal-fired power plants also were shut down (Ma 2014).

China's fuel-economy standards for automobiles were first established in 2004 under the formal authority of the General Administration of Quality

Supervision, Inspection, and Quarantine (GAQSIQ, 质检总局) with the cooperation of the former State Economic and Trade Commission (经贸委) and the NDRC (发改委), and they resulted in substantial fuel savings and reduced GHG emissions over time (Oliver et al. 2009). The cumulative amount of saved fuel from 2003 to 2012 is estimated at 13.8 million tons, and the avoided CO_2 emissions were 45.4 million tons. The fuel saved was equivalent to about 11 percent of total fuel use in 2010 (Jin et al. 2015).

Coal-consumption caps are another good example of regulatory/administrative policies in China and were frequently cited as one of the most effective policies to reduce GHG emissions in a survey of experts (Gallagher, Zhang, and Orvis 2018). The Airborne Pollution Prevention and Control Action Plan unveiled by the State Council in 2013 pledged to cap coal consumption and improve the air quality of the entire country by 2017. Thereafter, six Chinese provinces established absolute coal-consumption-reduction targets in their air pollution action plans, with a 50 percent reduction targeted in Beijing (北京), 13 percent in Hebei (河北), 19 percent in Tianjin (天津), 5 percent in Shandong (山东), 21 percent in Chongqing (重庆), and 13 percent in Shanxi (山西) by the end of 2017 compared to 2012 levels.

Reforms in the power sector could be considered either regulatory/administrative or fiscal policies. Generation and distribution were separated from each other to create more competition among generators to produce a competitive market price and to allow for new firms to enter the electricity market. The grid itself also needs upgrading so that it can accept more renewable energy generation, and this has been an ongoing process. Pilot projects in Yunnan (云南) and Guizhou (贵州) are being conducted. In 2015, the Chinese government announced a new *green dispatch* policy, but implementation policies are needed in the context of power sector reform so that nonfossil sources are prioritized on the grid.

Fiscal   The central government often couples fiscal policies with regulatory/administrative ones to create a more consistent and effective policy regime to incentivize behavior. It has, for example, subsidized the purchase of more efficient industrial equipment, the purchase of more fuel-efficient vehicles, and the deployment of both nuclear and renewable energy power plants. One of the most effective policies—at least at first—was the establishment of feed-in tariffs for renewable energy in China starting in 2003.

The feed-in tariff was coupled with a renewable energy development fund that came from a surcharge on consumption of coal-fired electricity. Developers of renewable energy would collect their feed-in tariff from this fund. In recent years, developers have been frustrated by long delays in receiving their payments, and the program has also suffered from gaps between the amount collected and the amount needed to pay out the subsidy. To partially address these problems (and the fact that the technological costs have fallen), the central government reduced the 2018 benchmark feed-in tariff for onshore wind power to RMB 0.40/kWh from RMB 0.57/kWh, representing a 15 percent cut from 2016 levels, and similarly reduced the FIT for grid-connected solar PV (see the appendix for more details; NDRC 2016).

Fiscal tools are also used to disincentivize behavior, such as the 2006 differential electricity tariffs that imposed a punitive electricity price on energy-intensive industries to encourage them to be more energy-efficient. Another power fiscal tool was the reduction of fossil fuel subsidies coupled with the imposition of a resource tax on crude oil and natural gas in 2011 and a new resource tax rate on coal in 2014. Between 2012 and 2014, fossil fuel subsidies were reduced from RMB 16.5 billion to RMB 11.8 billion in China (IEA 2015).

Industrial By some accounts, the Chinese government's efforts to reform and restructure the economy may have reduced GHG emissions more than any other policy measure because as the Chinese economy transitions away from heavy, energy-intensive industry and toward a lighter, service-based economy, energy consumption and GHG emissions naturally come down.

Many specific policy tools have been employed in the effort to make this transition, but the reform process has been among the most intractably difficult for Chinese policymakers because so many firms are state-owned enterprises employing thousands or even hundreds of thousands of workers, and local protectionism is fierce (Zhuang 2009). For locally state-owned enterprises, the local government is nearly always unwilling to shut down a factory that contributes to local GDP, provides wages for local workers, and contributes taxes to the local government. In addition, corrupt practices slow down efforts to reform these enterprises.

To optimize the industrialized structure, four types of policies were promulgated: (1) accelerating the elimination of backward production capacity, (2) promoting the upgrading of traditional industries, (3) supporting the development of strategic emerging industries, and (4) accelerating the development of the service sector. All four of these policies were included in "China's Policies and Actions for Addressing Climate Change" issued in 2016. Green finance policies including green bonds, green credit, and green rating systems were also established.

As arduous as it has been, the composition of the economy has changed dramatically since 1978, the year of China's opening up. In 1978, primary industry (i.e., agriculture, forestry, fisheries) accounted for 28 percent, secondary industry (i.e., manufacturing, mining, energy production, and construction) accounted for 48 percent, and tertiary (i.e., services) just 25 percent of GDP. As of 2015, primary industry contributed just 8 percent, secondary 41 percent, and tertiary 50 percent (National Bureau of Statistics of China 2016). In the thirteenth FYP, green development is put forth as a strategic objective, and the government launched thirty-nine major construction projects in the name of green development (Hu 2016).

Innovation   The Chinese government has long prioritized investment into energy innovation, but climate change was not a major driver of energy innovation policy until recently. After the reform and opening, the focus was mainly on acquiring, localizing, and deploying more advanced coal-combustion technologies (Zhao and Gallagher 2007), and China is now considered to have some of the most advanced coal technology capabilities in the world (Gallagher 2014). Relatively neglected has been a real effort to advance carbon capture and storage technology and gas turbine technology in China (Gallagher 2014; Liu and Gallagher 2010). The Ministry of Science and Technology has also placed increasing emphasis on innovation in "new-energy" vehicles, which originally encompassed hybrid electric, battery, and fuel-cell vehicles but now is overwhelmingly focused on pure battery-electric vehicles.

Although the Chinese government provided some early support for research and development of renewable energy technologies, it was surprisingly small. The Chinese government's main contribution to the development of renewable energy capabilities occurred through its strong industrial and fiscal policy supports to the nascent but rapidly growing renewable energy industry after 1995 (Lewis 2013; Lewis and Wiser 2007; Ru et al. 2012). Indeed, the central government now endeavors to foster a low-carbon economy through innovation policy, industrial policy, and market-pull incentives (Gallagher 2014; Zhang and Gallagher 2016). Innovation policy for nuclear energy technology has been very strong for fusion but weaker for fission. Although China has mastered generation II reactor technology, it relies on imports for more advanced nuclear technology (Zhou and Zhang 2010).

Experimenting with Explicit Climate Policy in the Twelfth and Thirteenth Five-Year Plans

Starting with the twelfth FYP, explicit domestic climate policies were pursued. Not surprisingly, the central government opted to experiment with these policies first, with the aim of expanding to the national level if the policies were successful. The central government invited local and provincial governments to apply to be low-carbon pilot regions. From the point of view of the local governments, the objective of participating was not only to reduce emissions but also to create new economic opportunities and hopefully obtain preferential treatment from the central government (e.g., approvals for acquisition of land) in exchange for their cooperation: the central government's Ministry of Land and Resources allocates the percentage of land allowed for urbanization.[3] Some local officials were genuinely enthusiastic about reducing GHG emissions, because they saw either economic opportunity in green development or cobenefits from reducing local air pollution. From the central government point of view, the objectives were to experiment with policy design and build support at the local level. After President Xi Jinping made the Joint Statement with President Obama, the Alliance of Peaking Pioneer Cities (APPC) was formed in China for those cities that would commit to peak years earlier than the 2030 national peak year. The idea was to stimulate competition among the local cities.

Market-Based Policies   Following the adoption of the Paris Agreement, a new approach began to be used in the context of China's explicit climate policies: market-based policies (Li et al. 2016). The central government recognized the need for market-based instruments because other policies were not as effective as they needed to be. A major question was whether to experiment with carbon taxes, cap-and-trade systems, or both. Bureaucratic politics strongly affected the outcome. The NDRC has control over climate policy but not taxation, so even though the Ministry of Finance (MOF) supported a carbon tax, and many within the NDRC supported a carbon tax, ultimately the NDRC did not support a tax because it wanted to be able to control the policy and could only do so if it was a cap-and-trade program. The NDRC won the fight with MOF for the first phase, but it may be that in the future, the ETS will convert to a carbon tax or that a carbon tax will be implemented in noncovered sectors under the ETS. The new emissions-trading system announced in 2017 initially covers the power sector and includes 1,700 enterprises and 3.3 billion tons of CO_2 (People's Daily 2017).

There is a long history of discussion about carbon taxes in China. At the start, there was worry about the potential impact on the economy of a carbon tax because economic growth was the main priority at the time. Therefore, for a long time, a carbon tax was just a policy option on the table that was not considered to be feasible to implement politically. Later, in 2016, an environmental tax was issued; however, though it covered a "basket" of pollutants, it did not include carbon. In general, the National Tax Bureau and the Ministry of Finance have both supported the idea of carbon taxes in China.

In 2011, the Chinese government initiated seven pilot carbon-emission-trading programs at the city or provincial level (Narassimhan et al. 2017; Qi and Zhang 2016). These pilots were required to launch by 2013 and fully initiate by 2015 (Zhang et al. 2014). The seven pilot programs were in Beijing, Tianjin, Shanghai, Chongqing, Shenzhen, Guangdong, and Hubei, and they only capped CO_2. The pilots' ETS designs varied widely and covered 7 percent of China's national emissions as of 2010. Nearly all of the permits were given away for free rather than auctioned, except for a small number that were auctioned in Guangdong, Shenzhen, and Hubei (Dong, Ma, and Sun 2016). As of August 2014, the average trading price across the seven pilots was RMB 6.7/ton CO_2eq (Yu and Lo 2015).

A number of problems were identified in the context of the pilot projects, including "poor GHG measuring and reporting practices, incomplete legal frameworks, noncompliance, ineffective enforcement, low penalties," and, in some pilots, illiquidity with low trading volumes (Yu and Lo 2015). The maximum allowable fine for noncompliance was only RMB 100,000 (USD 14,459 at 2017 exchange rates) per enterprise, which is not seen as an effective deterrent (Zhang et al. 2014). Most of the pilot ETS programs imposed far smaller noncompliance fees (Dong, Ma, and Sun 2016).

Significantly, a survey of Chinese firms conducted in 2015 revealed that the carbon price failed to "stimulate companies to upgrade mitigation technologies" and that the majority of firms considered participation in the ETS pilots as a means of improving ties with governments and earning a good social reputation rather than as a cost-effective mechanism to mitigate greenhouse gas emissions (Yang et al. 2016).

After identifying the potential and challenges of the pilot ETS programs, the central government commenced work on a "national" ETS that would cover some sectors in its initial phase beginning in 2017 and take into account lessons learned with the pilot programs.

Lessons Learned from Chinese Climate Policy

Several lessons can be learned from this examination of how climate policies have been implemented in China. First, there has been a steady strengthening of efforts to improve energy efficiency, reduce carbon intensity, and expand nonfossil sources of energy supply dating well back to the 1990s. Although the central government relied initially on indirect policies with multiple cobenefits, such as reduced conventional air pollution and improved energy security, it has increasingly moved to directly control GHGs.

Second, China's target allocation system is essential to the implementation of climate policies. The central government establishes the targets, but the subnational levels of government are those that must implement the policies to achieve them. At times, subnational governments have competing interests that compromise their willingness to fully implement the central government's targets.

Third, China's domestic climate policies are stable and comprehensive, encompassing almost all sectors. The central government tends to rely most heavily on regulatory and fiscal policy tools but increasingly has been experimenting with market-based policy instruments such as emissions trading.

6 Why Climate Policy Outcomes Differ

Now that we have examined the governance systems in each country, as well as case studies showing how climate policy targets are formed and implemented, we can explain why policy outcomes differ. We do not aim to be comprehensive in this chapter, but rather to identify some of the key differences that strongly influence policy outcomes. We conclude that, on balance, these factors best explain the differences in the policy processes of both countries.

We identify seven factors that differentiate the policy process in the two countries, and they can be grouped into three categories: political, economic, and social (as depicted in figure 6.1). Within the political category, we include party politics, separation of powers, government hierarchy, and bureaucratic authorities. Within the economic category, we include economic structure and strategic industries. The social category includes individual leadership and the role of the media. Of course, all of these factors are connected to one another and cannot be completely separated into these three categories. Individual leadership could, for example, be included in the political category, and the political economy of climate change encompasses all three categories.

Party Politics

Party politics are ubiquitous in both countries, but they manifest themselves so differently that they are almost incomparable. In the United States, the politics of environmental protection have changed so dramatically during the last forty years that what used to be an issue with broad bipartisan consensus has now become one of the most divisive topics today. In China, there is broad political consensus about the need to protect the

Political	Economic	Social
Party politics Separation of powers Government hierarchy Bureaucratic authorities	Economic structure and strategic industries	Individual leadership The role of the media

Figure 6.1
The main factors that cause the US and Chinese policymaking processes to differ.

environment and, in fact, reducing air pollution is one of the political imperatives most strongly felt by China's government leaders to maintain their legitimacy. The direct outcome of the political impasse on climate change in the United States has been the inability of the US Congress to pass comprehensive legislation on climate change even when a supermajority of Americans support it (Popovich, Schwartz, and Scholossburg 2017). The United States still has a long way to go in resolving political differences. The direct outcome of the political consensus in China is the inclusion of explicit climate policy in the thirteenth five-year plan and the decision to implement a national emissions-trading system in 2017. Despite its political consensus, China also has a long way to go in national law-making on climate change if it is actually to achieve a peak and then bend its emissions curve downward.

Historically, environmental conservation in the United States was associated with Republican presidents. Six years into his presidency, Republican Theodore Roosevelt (1901–1909) signed into law "An Act for the Preservation of American Antiquities," which allowed him and future presidents to proclaim national monuments without resort to Congress. He created five national parks, designated the Grand Canyon a national monument, and proclaimed fifteen other national monuments, thirteen new national forests, and sixteen national bird refuges (Morris 2001). Republican Richard Nixon (1969–1974) signed into law the Clean Air Act and the Clean Water Act, and Republican George H. W. Bush (1989–1993) signed the Clean Air Act Amendments of 1990 and supported the ratification of the UN Framework Convention on Climate Change.

During the Bill Clinton presidency (1993–2001), most Republicans in Congress turned against the environment. The timing of this change was ironic given that Bill Clinton's vice president, Al Gore, was one of the first politicians to champion the need to act on climate change. As described in chapter 3, the Republican "revolution" led by Congressman Newt Gingrich was the turning point at which the Republican party turned away from environmental protection and displayed particular hostility toward global environmental challenges such as stratospheric ozone depletion and climate change. It is important to emphasize that individual Republicans have championed climate change policy since then, but they have been lonely. The Waxman-Markey bill that passed the House was led by Democrats, and the equivalent bill could not pass the Senate even though two prominent Republicans, Senators Lindsey Graham and John McCain, put their support behind it. Although most Democrats support climate policy, a small minority of Democrats from coal-mining states have consistently opposed climate legislation—including Senator Joe Manchin from West Virginia, who infamously used a rifle to literally shoot a hole in a paper copy of the cap-and-trade bill in a campaign TV advertisement.[1]

The Republican Party has also become the party of the fossil fuel industry in America. Campaign contributions from the fossil fuel industry have become highly skewed toward Republicans. Since 1990, more than two-thirds of oil and gas campaign contributions went to Republican candidates. The mining sector's contributions were even more skewed toward Republicans, with 90 percent of such donations going to Republican candidates (Center for Responsive Politics n.d.).

The situation could not be more different in China where climate change is not a political issue in the same way it is in the United States. There is little public debate about the science of climate change among government leaders there, and little debate in the public about the need to reduce GHG emissions (perhaps because the state controls the media). Disagreements can be observed in social media, but the main discussion is not whether to act but how *much* to do, where, and how to do it. These disagreements are not expressed in political terms; in other words, attitudes on climate change—for or against—are not connected to political parties in China. Even though the Communist Party of China (CCP) dominates the political sphere, other parties do exist, as discussed in chapter 3, but they are weak and it is not possible to differentiate them based on environmental issues. Within the CCP, general consensus on the need to act on climate change has been achieved at the highest level—the Politburo Standing

Committee—and this is shown by the ratification of the Paris Agreement and the inclusion of carbon-intensity targets in the five-year plans because Politburo approval is needed to take these steps. To the extent that there is a politics of climate change in China, it manifests in bureaucratic struggles among the ministries associated with different industries and debate within the party about how much burden to put on the fossil fuel industries and energy-intensive enterprises—the revenues of which, after all, are needed to support the state and to maintain employment.

The politics of climate change in China thus is completely different from in the United States. In China, energy-intensive enterprises that believe they would suffer from climate policy will try to use the power of local governments (whose tax revenue and local employment rely on these enterprises) to resist central government policies. The main struggles are thus not among the parties, but among and within the different levels of governments in China.

Separation of Powers

An idea that is cherished in the United States is the *separation of powers*, or the idea of distributing governing power across three separate branches of government: executive, legislative, and judiciary. In *The Federalist Papers*, a collection of essays written to explain and defend the US Constitution, James Madison argues that "the accumulation of all powers, legislative, executive, and judiciary, in the same hands, whether of one, a few, or many, and whether hereditary, self-appointed, or elective, may justly be pronounced the very definition of tyranny." He acknowledges, however, that these branches cannot be totally separate and distinct from each other in the practical operation of a government (Hamilton, Madison, and Jay 1961, 301–302). The main goal of the separation of powers in the US Constitution is to introduce checks and balances so that no one branch or powerful person can exercise unchecked power.

The separation of powers in the United States is not mirrored in China (Li, Li, and Qi 2011). A strong administrative state dominates over the legislative and judicial branches of government, and the CCP exerts control throughout all three branches by infusing them with CCP leaders and members. Although China has a National People's Congress, and it was legally established in the first Constitution of the People's Republic of China in 1954, it is demonstrably weaker than the administrative branch. Although the NPC used to be known as a "rubber stamp," Chinese experts believe that view is now outdated (Shi 2014). China's judiciary is also weaker than the administrative branch; according to one account, "the spread of judicial corruption has weakened and undermined public trust in and support of the Constitution, the legal system, the party, and the government" (He 2014, 373). Still, the judiciary in China is professionalizing and even has created a specialized environmental court system (Wang and Jie 2010).

The CCP is the organizer, manager, and coordinator of Chinese society. Zheng Yongnian (2010) coined the term *organizational emperor* to elucidate the CCP's role in Chinese society. The top leader of the CCP plays a role similar to that of a patriarch in a large family. In Chinese traditional culture, the family and the state are essentially the same (家国一体): the state is like a big family, and the family is like a small state. This political structure has a profound influence on policy in China, and the principle of separation of powers has both weak legitimacy and weak implementation in the Chinese context. Instead, power is totally concentrated in the hands of CCP leadership.

The separation of powers in the United States has strongly affected climate policy outcomes there. The US Senate's refusal to ratify the 1997 Kyoto Protocol and the 2015 Paris Agreement on Climate Change suppressed domestic climate legislation that would have worked toward fulfilling these these international agreements. Despite President Obama's strong support for a domestic climate bill that was not necessarily tied to the international negotiations, the US Congress was unwilling to pass comprehensive climate legislation during his presidency. As a result, President Obama's options were constrained primarily to use of his executive authorities under existing laws, which was why he chose to pursue a regulatory approach to limit GHG emissions during his second term. When previous US presidents opposed climate policy, interest groups and US states used the judiciary to advance climate policy. The Supreme Court decision that the EPA was obligated to regulate CO_2 and other GHGs under the Clean Air Act during the George H. W. Bush administration was a good example of the separation of powers creating checks and balances in the system. In President Trump's administration, attempts by EPA Administrator Scott Pruitt to roll back the regulations issued by President Obama are being litigated in the courts and may ultimately prove unsuccessful.

The relative lack of separation of powers in China has been consequential in a completely different way for climate policy there. China's collective decision-making model requires a consensus-based approach to new domestic climate policies, and most of the new climate policies are developed and implemented by ministries within the executive or administrative branch of China's government. The National People's Congress, however, passed several foundational laws relevant to climate change, including the 2005 Renewable Energy Law, the Energy Conservation Law, Circular Economy Promotion Law, and the Electricity Law. The Renewable Energy Law laid the foundation for clean energy development in China. Because of the centralization of the CCP and lack of separation of powers, it is much easier in China to design more coherent packages of policies that are aligned, consistent, and complementary with each other. The fiscal subsidies, regulations, and industrial policy in support of renewable energy, for example, are all aligned to promote the rapid deployment of renewable energy to meet the ambitious nonfossil energy targets in the thirteenth FYP. The administrative branch and the NPC thus play reinforcing roles for each other in the Chinese context.

On the other hand, implementation and enforcement of climate policies are weakened by the absence of separation of powers in China. In the United States, any individual or organization that believes it is being harmed by a firm that is not complying with federal environmental laws or regulations can sue the firm in a federal court, but in China it is much less common and more difficult to do so. In China, the central government relies upon lower levels of government to enforce its policies (see next section), and the lower levels of government are sometimes reluctant to compel behavior that harms local firms that either are owned by the local government itself or provide helpful tax revenues and employment to the locality. In addition, these local officials are frequently rotated and have short time horizons as a result (Eaton and Kostka 2014). Because judicial appointments lack independence from the party, recourse to the courts is much less of an option for citizens seeking remedy from harm. As a result, the Chinese system "is more effective in producing [environmental] policy outputs than outcomes" (Gilley 2012).

Government Hierarchy

The structure of government in terms of its hierarchy is one of the greatest differences between the policymaking systems of the United States and

China. As discussed in chapter 3, China has a hierarchical, pyramidal structure of government in which the system is vertically oriented through five tiers: central (中央), provincial (省), prefecture (市), county (县), and town or village (乡镇). Each level may create policy, but all levels are obligated to implement and enforce policies issued by higher levels, and all lower levels must obtain the approval of higher levels before issuing important new policies. As discussed in the previous section, local governments sometimes lack sufficient incentives to implement and enforce central government policies.

In the United States, the policymaking processes of the federal and state governments function in parallel, and the Tenth Amendment to the US Constitution states that the "powers not delegated to the United States by the Constitution, nor prohibited by it to the States, are reserved to the States respectively, or to the people." In other words, states are empowered to enact new policies without the approval of the federal government so long as they are not contradictory to federal laws or the Constitution itself. Importantly, this means that the federal government enforces its own policies, and state and local governments enforce their own policies. This crucial difference in the responsibilities for enforcement of power between the United States and China leads to quite different policy outcomes.

For climate policy, US states are fully empowered to pass and enforce their own laws on climate change unless prohibited by federal law. The federal Clean Air Act empowers the Environmental Protection Agency to set national standards for air pollutants (including GHGs), and no state can have a weaker standard. To limit the amount of variation in stricter standards that industry must adhere to, the CAA authorized California to set more stringent standards because it had already done so when the CAA was passed, and other states are allowed to adopt California's standards. The states of Massachusetts and New York routinely adopt California's environmental standards, for example. Some states have passed new and different types of laws to achieve greater reductions in GHGs. The northeastern states and California both created cap-and-trade programs to reduce emissions. More than thirty-five states have passed laws that create renewable energy portfolio standards. As a result of the authority granted to states, they can individually or collectively be much more progressive than federal policies. And, in the absence of federal policy, they can lead the way (or, conversely, refuse to act).

China's hierarchical system means that the provinces are very unlikely to pass laws that have not received prior approval from the central government. In theory, based on Article 63 of the Legislation Law, provinces have the authority to enact local decrees (地方性法规) provided that they do not contravene any provisions of the Constitution or national law. But, in reality, provinces would not pass a law that would be disliked by the central government given that the top leaders of the provinces are appointed by the central government or the CCP. In climate policy, the provinces and localities are strongly encouraged by the central government to take on more aggressive targets and implementing policies. The Association of Peaking Pioneer Cities (APPC), which was guided by the central government, provides a nice example of how the central government can encourage provinces to adopt more aggressive targets than the national one. The APPC secretariat is based at the Academy of Macroeconomic Research, which is affiliated with the central government's NDRC (APPC 2017).

Individual Leadership

Leadership is a necessary but not sufficient condition for climate policymaking in both the United States and China. Leaders are required both at the highest levels and at senior levels in the bureaucracy.

Although it is not possible to say for sure, it is unlikely that the 2014 Joint Announcement on Climate Change would have been made without risk-taking and personal leadership from both Presidents Xi and Obama. Relatedly, without senior leadership from individuals like Counselor John Podesta, Science Advisor John Holdren, Secretary John Kerry, Secretary Ernie Moniz, and Special Envoy for Climate Change Todd Stern on the US side and State Councilor Yang Jiechi (杨洁篪), Vice Premier Zhang Gaoli (张高丽), Minister Xie Zhenhua (解振华), and Minister Wan Gang on the Chinese side, and key members of their staffs, it is hard to imagine that this agreement could have come into fruition.

The reason that leadership is not sufficient is that other factors in the policymaking process can outweigh its influence. One example from the United States is that Vice President Al Gore was unable to achieve either of his primary objectives related to climate change during the Clinton-Gore administration—namely, the imposition of a BTU tax (fuel tax) and Senate ratification of the Kyoto Protocol to the UNFCCC. Few American leaders are more committed than Al Gore to climate change policy, but even as vice president of the United States he could not advance it as much as he would have liked.

In the two climate change bills that came closest to passing Congress, Congressmen Henry Waxman and Edward Markey were key to the development and passage of the bill in the House. Similarly, Senators Joseph Lieberman, John McCain, and Lindsey Graham were the leaders of the Senate effort, even though it was not ultimately successful. President Obama's remarkable advances in regulatory policy under his Climate Action Plan came about in recognition of the fact that he had existing authority under existing laws that he could use to reduce GHG emissions, and his EPA administrator in his second term, Gina McCarthy, was willing to lead the efforts to develop and implement all of the various rules that would be required. Secretary of Energy Ernest Moniz led many efforts on innovation and energy-efficiency policy as well.

On the Chinese side, individual leadership is harder to discern because the policy development occurs through a behind-the-scenes, consensus-seeking process, but many Chinese experts credit Minister Xie Zhenhua (解振华) as one of the driving forces behind both domestic and international climate policy in China. The former leader of the predecessor to the Ministry of Environmental Protection, Minister Xie became the vice chairman (ministerial rank) of the National Development and Reform Commission in 2006. The Department of Climate Change (应对气候变化司) under the NDRC reported to him until he stepped down as vice minister in 2015, although he remained the special representative for climate change through the negotiation of the Paris Agreement. His long-time deputy—the former director general of the climate department in the NDRC, Su Wei—was responsible for the formulation of many domestic climate policies. Similarly, Minister of Science and Technology Wan Gang is credited for being a strong force for clean energy innovation in China, in particular for advanced clean vehicles.

Economic Structure and Strategic Industries

A country's economic structure and industry composition create policy constraints and opportunities because industries will either advocate for or resist new climate policies. In the United States, the economic power of the fossil fuel industries dwarfs that of the renewable energy industries, and their campaign contributions to politicians reflect this imbalance. In 2016, total campaign contributions from the wind and solar industries were USD 1.4 million compared with USD 121.2 million from the oil, gas, and coal industries (Center for Responsive Politics n.d.). Still, industry advocacy coalitions have been formed by renewable energy producers in the United States, such as the American Wind Energy Association (AWEA) and the Solar Energy Industries Association (SEIA), to voice their interests in Washington, DC.

Even within some of the industrial interest groups, divisions can occur. A split in the US solar industry occurred, for example, when some US-based manufacturers, led by German company Solar World, asked the US government to impose tariffs on Chinese PV manufacturers because they believed the Chinese firms were unfairly subsidized. Solar panel equipment marketing and installation companies that believed import tariffs on Chinese equipment manufacturers would cause harm to US installation jobs if the imports cost more formed a collation called the Coalition for Affordable Solar Energy (CASE). According to CASE, the US solar industry "provides jobs for over 142,000 Americans—and nearly 80 percent of those jobs are in solar installation, project development, and sales."[2] The other industry coalition, the Coalition for American Solar Manufacturing (CASM), applauded the US government's decision to impose tariffs on Chinese producers.[3] All of the solar industry coalitions supported the extension of the solar investment tax credit, however.

In China, there are three fundamental factors affecting how economic structure and strategic industries have influence on policy outcomes. First, China is still in the midst of an industrialization process, even though the share of tertiary industry has recently surpassed that of the secondary industry in GDP. The heavy industries such as iron, steel, chemicals, and cement still play very important roles in the economy. Second, China's natural energy endowments are overwhelmingly dominated by coal, although China also has good natural renewable energy resources. Coal still accounts for more than 60 percent of energy supply, so the coal industry will have a significant influence on policymakers for years to come. Third, many of the large heavy industry companies are state-owned enterprises, some of which are owned by the central government and some of which are owned by provincial or local governments. The large SOEs owned by the central government will use their special political position to try to influence policy in different ministries, especially the NDRC and MIIT. Locally owned SOEs will be even more influential in some cases, such as in coal-mining regions, because these local SOEs have so many employees that the local governments will be very cautious about making and implementing policies that will harm them.

Bureaucratic Authorities

Climate change policy is managed quite differently in China compared with the United States. The bureaucracies that engage on climate change policy do not mirror each other in the two countries, and this leads to confusion on both sides about who is responsible for which aspects.

In China, the ministry that had primary authority over climate policy until 2018 was the National Development and Reform Commission, the former State Planning Commission. This ministry has many complex functions, but the ones that were most germane to climate change are formulating and implementing strategies for national economic development; creating annual plans; directing, promoting, and coordinating the restructuring of the economic system; formulating comprehensive industrial policies; coordinating energy saving and GHG emission reduction; and organizing "the formulation of key strategies, plans and policies in addressing climate change," including taking the lead "with related ministries" in attending international negotiations on climate change and undertaking relevant work in regard to the fulfillment of the United Nations Framework Convention on Climate Change at the national level (NDRC 2017a). The National Energy Administration, which is responsible for formulating and implementing plans for the energy industry, is part of the NDRC.

In 2018, as discussed in chapter 4, the government announced a reform and created a new Ministry of Ecology and Environment (MEE) (生态环境部), replacing the Ministry of Environmental Protection and moving the responsibility for climate change policy from the NDRC to the new ministry (Xinhua 2018). The NDRC retains responsibility for China's industrial and energy policies, however, which means that most of the indirect climate policies will continue to be managed by the NDRC. The new MEE retains responsibility for regulating conventional air pollution, as well as some of the non-CO_2 GHG emissions, especially those that are governed by the Montreal Protocol on Substances that Deplete the Ozone Layer, such as hydrofluorocarbons (HFCs).

Another new ministry announced in 2018 is the Ministry of Natural Resources (MNR; 自然资源部), which will include the State Forestry Administration, renamed the State Forestry and Grassland Administration. The new MNR will manages all forestry and land-use policies, and these sectors are important sources and sinks of greenhouse gas emissions. Other ministries with strong direct or indirect climate change responsibilities include the Ministry of Industry and Information Technology (MIIT), Ministry of Finance (MOF), Ministry of Science and Technology (MOST), Ministry of Commerce (MOFCOM), and Ministry of Foreign Affairs (MOFA). MIIT has responsibility for many industrial policies, including those supporting strategic emerging industries like clean energy. MOF has responsibility for taxation, so it develops and implements the environment and resource taxes. MOST is responsible for promoting innovation and investing in R&D in the energy industry. MOFCOM is responsible for formulating and implementing domestic and foreign trade and international economic cooperation. MOFA normally takes the lead on international negotiations, except in the case of climate change, where the NDRC has traditionally led, although MOFA joins the international climate negotiations and has its own climate change office in the Department of Treaty and Law.

The most prominent example of how the bureaucratic arrangements affect policy outcomes in China was the struggle between the State Administration of Taxation and the NDRC over the choice of carbon-pricing policy instruments. The State Administration of Taxation was affiliated with MOF, and both have long proposed the development and implementation of a carbon tax, but the NDRC opposed it in part because it would not have authority such a tax. The NDRC instead advocated for a cap-and-trade program. The NDRC won the bureaucratic battle in 2015 when President Xi Jinping announced that China would launch a national ETS by 2017. Also, the fact that the NDRC leads the Chinese delegation to the international climate negotiations means that China's domestic and international climate policies are managed by the same office, which makes it more likely that China's international commitments will be honored domestically and vice versa.

In the United States, the US State Department leads the delegations to the international climate negotiations. Traditionally, the Bureau of Oceans and International Environmental and Scientific Affairs (OES) had responsibility for climate negotiations, but under the Obama administration, a special envoy for climate change (SECC) was named, with his own office. The National Security Council (NSC) in the White House is supposed to coordinate policy and strategies among agencies related to international affairs, and under the Obama administration there was a senior director for energy and climate change. Tensions typically emerge between the State Department and the NSC about who is responsible for which aspects of foreign policy.

Domestically, the US Department of Energy (DOE), Environmental Protection Agency (EPA), US Department of Transportation, Department of the Interior (DOI), Department of Agriculture (USDA), National Aeronautic and Space Administration (NASA), and National Oceanographic and Atmospheric Administration (NOAA) are the primary agencies responsible for climate policy. DOE has the authority to design and implement energy-efficiency standards. DOT has the authority, together with EPA, to design and implement fuel-economy standards for motor vehicles. EPA has authorities under the Clean Air Act and Clean Water Act to regulate GHG emissions. DOI and USDA have authorities to manage land-use change and forestry emissions. NOAA and NASA are the two main science agencies that provide data and information regarding climate science and emissions trends.

The arrangement of these authorities in the US federal government means that sometimes certain agencies may take steps that are contrary to the objectives of others. In 2015, for example, after President Obama had announced the US climate target, the Bureau of Land Management (BLM) within the DOI issued a plan to lease more federally owned land for coal production. Increased coal production was clearly at odds with the US climate target, so six months later the Obama administration halted federal coal leasing to reassess its policy in a comprehensive three-year review (Kolbert 2015; Magill 2016).

The Role of the Media

The influence of the media on the policymaking process is fundamentally different in the United States and China because of CCP control of the media and Internet in China and the lack of any equivalent control in the

United States. Although most newspapers and media outlets operate on a commercial basis in China, their content is supervised by the Propaganda Department of the CCP (宣传部; Shirk 2011) and other related government bodies such as the Office of the Central Leading Group for Cyber Affairs (中共中央网络信息安全和信息化领导小组办公室 or 网信办), which focuses on the supervision of the Internet, and the State Administration of Press, Publication, Radio, Film, and Television (新闻出版广电总局or 广电总局), which supervises books, newspapers, radio, film, and television (Yang 2017). If an article is published that does not meet the approval of the CCP, then it will be taken down. Likewise, a post on a blog or a webpage on the Internet can be banned. Many Western news sources are blocked in China, including the "newspaper of record" in the United States, the *New York Times*. Lots of Western social media apps and search engines are also banned or limited in China. Prohibited content includes anything regarded to be harmful for the stability of CCP rule.

In China, it is believed that the top leaders of the country have a moral responsibility to educate its people, based on traditional Confucian culture in which a father must take responsibility for educating his child. In addition, the CCP regards the media as a propaganda tool that is necessary to control in order to maintain its rule, similar to the situation with the military. The growth of civil society and interest groups has led to expanded individual consciousness, but it is strongly influenced and suppressed very seriously in some cases.

Chinese media coverage about climate change itself, and climate change policy specifically, appears to be relatively open. The government encourages coverage of the science of climate change and its impacts because such articles and videos help to educate the public about the need to act, and this is regarded as helpful for the party and the government. As a result, there are many articles and media reports that explain, advocate, and even propagandize climate policies to the public. A 2017 survey of Chinese adults found that more exposure to media content increased support for an international climate treaty (Jamelske et al. 2017). Even opinion articles critical of policy details are permitted, such as one published in *Low Carbon World* on the problems of the carbon-trade pilot programs in China (Yin 2015), so long as the overall direction of the policy is not undermined. The government also encourages climate coverage to educate the public and encourage individuals to adopt a low-carbon lifestyle.

The extent to which the Chinese government intervenes in the media to censor content on climate change is not well studied, but it does not seem that the government permits widespread skepticism of the science itself because such articles rarely appear in mainstream media—although some do exist, as described in chapter 1. However, within social media apps like WeChat, for example, lots of skeptical arguments are raised by individuals or in group chats, and fierce disputes take place without heavy censorship.

By comparison, the US government does not control or censor US media companies, although certain practices are prohibited by law (e.g., child pornography; US Justice Department 2018). Free speech and Internet privacy is expected by the American public and fiercely defended by public-interest groups such as the American Civil Liberties Union (ACLU) and the Electronic Frontier Foundation (EFF), although some laws were passed after 9/11 that allow the government to violate citizen privacy when the government has national security concerns; these laws remain controversial.

The downside to free speech generally prevailing in America is that misinformation and conspiracy theories are allowed to fester. The spread of misinformation or "alternative facts" occurs throughout all conventional media sources, social media, and websites. Climate-skeptic websites abound, such as https://junkscience.com/ and https://www.cato.org/research/global-warming, and are countered by websites run by environmental NGOs or climate scientists, such as http://realclimate.org or https://www.skepticalscience.com/. After the 2016 election of President Trump, many citizens called on Facebook and Twitter to address the spread of falsehoods and misinformation, and in 2017 Facebook announced that it would boost posts that are "genuine, not misleading, sensational or spammy" to try to limit fake news (Kulp 2017).

In the United States, the media contributed to public confusion about climate change during the 1990s and 2000s, and this often occurred through journalistic attempts to present "balanced" stories—that is, reporters providing different opinions that are given equal weight even if they are not representative of the majority view of scientists (Boykoff 2007). After the Fifth Assessment Report of the Intergovernmental Panel on Climate Change was released in 2013, the American news media usually presented the mainstream scientific consensus in shorter, summary form without offering a critique of it (Russell 2013). In other instances, questioning of science in the media was pursued intentionally, as documented in the book

Merchants of Doubt (Oreskes and Conway 2010), which described contrarian scientists, some directly funded by the fossil fuel industry, working assiduously to cast doubt on mainstream climate science through creation of websites and by being quoted in public media outlets. These contrarians frequently were the so-called experts quoted in the stories, providing a counter point of view even though they had not done any peer-reviewed research on climate change. Although the influence of this small handful of skeptics on the media has diminished considerably in the United States, it appears that they had a lasting influence on the American public given the greater levels of skepticism that exist in the United States than elsewhere in the world.

In conclusion, these seven differences—party politics, separation of powers, government hierarchy, individual leadership, economic structure and strategic industries, bureaucratic authorities, and the role of the media—are far from being the only ones that influence policy outcomes in the United States and China. They are, however, key to understanding how and why policy outcomes differ in the two countries.

7 Conclusion

As the titans of the climate, the actions of the United States and China are intrinsically essential to confronting the climate-change challenge during the twenty-first century. As the two largest aggregate emitters, these two countries must sharply reduce their own emissions by mid-century to avoid dangerous impacts from climate change. Moreover, US and Chinese leadership will be required again and again to create the conditions for new and improved international agreements in the decades ahead because climate change is truly a global phenomenon. Their leadership will only be accepted internationally if they are effective at reducing their own emissions at home. For both of these reasons, we focus primarily on domestic policy in this book.

Similarities and Differences

We have now thoroughly explored how the United States and China develop and implement climate-change policy domestically. We have identified key differences and a few similarities in the two countries' policymaking systems. What should be recognizable to people from both countries is that bureaucratic politics exist everywhere. Bureaucrats constantly negotiate with each other to resolve policy differences. They jockey with each other for influence with superiors, strive for promotion, and seek to have their views prevail in policy decisions. The authorities of different ministries, departments, agencies, and bureaus within each government create a structure for the policy process because these government entities have particular authorities over certain aspects of policy. Also, individual leadership can be important even in systems with collective decision-making because there is a need for determined champions of policies, sometimes called *policy entrepreneurs* in the United States. These policy entrepreneurs identify a policy solution, try to convince the leadership of their own ministry or department, and then work to sell it to counterparts in other ministries or departments to build consensus around the new policy.

There are also fundamental differences between the United States and China that derive from their constitutions, history, politics, and culture. One-party rule in China eliminates much of the policy gridlock that is experienced at the federal level in the United States with the mutual stubbornness of the Republican and Democratic parties' determination not to yield to each other. On the other hand, the lack of structural checks and balances on the Communist Party renders it vulnerable to making serious mistakes. Similarly, the constitutionally mandated separation of powers in the United States makes it very difficult for the president to take steps that may be illegal or unwise. In China, there are few restraints on the power of the seven to nine members of the Politburo Standing Committee, although they may feel pressure from other countries, feel pressure from domestic voices, or be influenced by each other. The Politburo Standing Committee sits at the very top of a hierarchy that reaches through five levels of government all the way down to the local level in China. The Chinese Communist Party pervades not only the administrative branch of government, but also the National People's Congress (NPC-全国人大), the Chinese People's Political Consultative Congress (CPPCC-全国政协), the courts (法院), and the military (军队). The central government is largely responsible for the formulation of policy, and subnational governments are responsible both for implementing the central government's policies and formulating and enforcing their own policies so long as they are approved by higher levels. In the United States, by contrast, the federal and state governments exist in parallel, each able to develop, promulgate, implement, and enforce their own policies, and they do so in three distinct branches of government at each level: executive, legislative, and judicial.

Finally, the economic structure of each country creates fundamentally different interests and motivations for their governments. The deindustrialization of America is currently one of the most serious political challenges for US politicians as a frustrated and disenfranchised American labor force struggles to cope with the disruptive forces of globalization and the changing nature of work. In China, the struggle is how to shift away from resource- and pollution-intensive industries and toward innovation-based,

technology-intensive, and green industries, all while sustaining employment for a population more than four times the size of the US population.
Central and local ownership of many companies is a distinct feature of
the Chinese economic landscape, and SOEs are a source of income to Chinese governments, especially at the subnational levels. By contrast, there
is virtually no government ownership of firms in the United States. One
similarity, however, is that US federal government expenditure accounts
for 20 percent of GDP, and Chinese government expenditure accounts for
25.7 percent of GDP in China, although most of the expenditure is at the
subnational level in China (Federal Reserve Bank of St. Louis 2018; National
Bureau of Statistics of China 2016, table 7-1).

Answering Legitimate Questions from Both Sides of the Pacific

In the introduction, we posed a number of questions that are answered in
considerable detail in the book. Here we provide simplified answers to each
one. First, why has climate policy been bottom-up in the United States and
top-down in China? The short answer is that China's political structure does
not allow for pure bottom-up approaches, although it does encourage the
piloting of experimental policies in the provinces. Any subnational reforms
or initiatives must be recognized and approved by the central government
before they can proceed. In the United States, even under determined leaders like Vice President Gore or President Obama, the federal government
has been unable to pass comprehensive climate legislation. But the states
can independently pursue their own climate change policies, and many are
doing so. The federal government has historically followed the states by
modeling new federal environmental regulations after ones that worked in
California or elsewhere.

Second, why has the United States not chosen economically efficient,
least-cost, market-based climate policies so far? In the United States, any
policy that would collect or redistribute revenue requires authorization by
Congress, which has the power of the purse. Climate change legislation
came close to passing in 2009, and the bill that passed the House of Representatives embraced a market-based cap-and-trade system for reducing
GHGs, but ultimately it did not pass the Senate. At the state level, however,
ten states have implemented emissions-trading systems, and many more
have renewable portfolio standards.

China has already experimented with seven pilot cap-and-trade programs at the subnational level, and the central government announced an emissions-trading system in 2015 that will cover a limited number of sectors when it launches in 2017 before gradually expanding to other sectors over time. In addition, the National People's Congress approved a National Environmental Tax Law in 2016, to go into effect January 1, 2018. At present, there is no formal "tax" on pollutants but instead there are "fees," according to the administrative regulations on levy and use of pollutant discharge fee that were issued in 2003 (State Council 2003). This tax does not include GHGs, but it will indirectly affect GHGs because other pollutants, derived from the same sources, are covered. Because China typically uses an administrative or regulatory approach, emissions trading will be a major departure from the past. So, why is China beginning to embrace emissions trading? Because the government seeks to limit the economic costs associated with reducing GHG emissions, and this was the NDRC's preferred approach given that it has authority for climate policy in the central government.

Third, why is the United States using a regulatory approach to climate policy even when it has historically claimed to prefer market-based approaches to environmental policy? As we now know, Congressional climate legislation failed in 2010, which is why President Obama decided to use his regulatory authorities under existing laws including the Clean Air Act to reduce GHG emissions from certain sectors. Many presidents, both Republicans and Democrats dating back to the 1970s, have utilized regulatory approaches such as energy-efficiency standards, which derive from authorities given to the relevant government agencies from various energy laws passed by Congress.

Fourth, how and why did both countries arrive at their climate change targets, and are they strong enough? We provided a comprehensive account of why each country chose its respective national target in chapter 4, but to summarize, the 2014 target-formation process was rather different in the two countries, with the White House leading a bottom-up interagency process in the United States, culminating with a decision by President Obama, and the NDRC coordinating a high-level interagency process in China, with considerable expert input, that culminated with a Politburo Standing Committee decision in China (with President Xi playing a dominant role in this decision-making process). Commentators in both countries criticized the other side as choosing weak or inadequate targets. The Senate majority leader Mitch McConnell said that the climate change agreement between the United States and China "requires the Chinese to do nothing at all for 16 years" (Jacobson 2014). Chinese commentators expressed concerns about the ability of the United States to live up to its commitment because Congress has, to date, never passed explicit climate change legislation. They also worried about the true nature of US intentions given that the US negotiators in Paris avoided the use of the word *shall* with respect to target-setting (Teng 2015; Zhao 2017). However, as we documented, both countries sought to find targets that were simultaneously ambitious enough to inspire the global community and realistically achievable so that both countries could feel confident they would responsibly honor their commitments.

The enforcement procedure is so profoundly unalike in China and the United States that it is unrecognizable to many people on both sides. In China, subnational governments are delegated the responsibility to implement and enforce higher-level policies. The provinces implement central government policies, and the municipalities implement both central and provincial policies, for example. The subnational governments are given some discretion in how they implement policies in acknowledgment of the wide variety of circumstances in individual provinces, counties, municipalities, and towns. Enforcement is primarily achieved through performance evaluation of subnational officials. There are two parallel performance-evaluation systems: one within the administrative branch of the central government and another in the CCP. The performance-evaluation system supervised by the organization department of the CCP focuses mainly on the officials themselves and is secret and not open in most cases; however, the performance-evaluation system of the administrative government mainly focuses on the region (not the individuals), and this process is open within the government. One provincial government will, for example, evaluate the GDP growth rate of different municipalities within the province. Failure to do a good job implementing the policies of his superiors will limit a local government official's chances of promotion and career advancement. The problem is that his interests may be conflicted by local government ownership of factories or by competing priorities, such as job creation and boosting GDP (Yang 2017).

The judicial system is rarely used as an enforcement procedure in China, but in the United States it is the primary means of enforcement. US federal officials can implement and enforce federal laws and regulations directly at the local level. They work in parallel with state and local governments. If a firm is not in compliance with a federal regulation, the federal government can file a civil or criminal lawsuit against the company to force the company to come into compliance or else face serious consequences.

The question of how institutional and bureaucratic differences affect climate policy outcomes is a big one, and it is answered throughout this book. One illustrative example is that up until 2018, the government agency with primary authority for climate policy in China was the National Development and Reform Commission (NDRC), which has no equivalent in the United States.[1] Responsibilities for climate change were transferred from the NDRC to a new Ministry for Ecology and Environment in 2018, which replaces the former Ministry of Environmental Protection. Since 1982, Chinese government institutional reform has occurred seven times, and each time the authorities and functions of the different government ministries have changed. A number of ministries besides the NDRC play specific roles in climate change policy in China. The NDRC historically focused on broader climate change policies, but in certain areas, other ministries would formulate more detailed policies for different sectors. The Ministry of Housing and Urban-Rural Development (MOHURD) will determine the policy for improving building energy performance, for example, and the Ministry of Industry and Information Technology will formulate motor vehicle fuel-economy standards. The Ministry of Science and Technology takes primary responsibility for energy innovation, the Ministry of Environmental Protection has authority for regulating non-CO_2 emissions, and the National Energy Administration has authority over China's energy industries.

The US government bureaucratic arrangement is completely different. US government agency authorities are directly derived from laws passed by the US Congress. Taking the example of implementing fuel-economy standards, the EPA has the authority to regulate CO_2 emissions from motor vehicles, and the National Highway and Transportation Safety Administration (NHTSA) of US DOT has the authority to regulate motor vehicle fuel efficiency, even though these areas overlap heavily. NHTSA's fuel-consumption standards are authorized under the Energy Independence and Security Act of 2007, and EPA's GHG emission standards are authorized under the Clean Air Act (EPA 2017c). In addition, the EPA and DOE work together to produce http://fueleconomy.gov, which provides information to consumers about the fuel economy of new vehicles.

Now, we turn to the final question we posed in the introduction: Why does China appear to welcome international agreements more readily than the United States? In fact, this question has only been answered indirectly thus far. In the United States, the US Senate was given the authority under the Constitution to ratify international treaties, but the president has the authority to negotiate executive agreements without Senate advice and consent. For climate change, although the US Senate ratified the UN Framework Convention on Climate Change (UNFCCC) in 1993, it has so far refused to ratify any of the subsequent international agreements related to the UNFCCC, most infamously the Kyoto Protocol. As a result, the US-China agreement was negotiated as an executive agreement, and the legally binding provisions of the Paris Agreement were formulated as decisions of the Conference of Parties of the UNFCCC. The US Senate's stance is emblematic of a view of America as preeminent and without equal in a unipolar world. In the minds of US political conservatives, the United Nations cannot get anything done and is no substitute for US leadership elsewhere in the world. The idea that the United States would be subservient to a supranational body like the United Nations is anathema to contemporary US foreign policy realists, and therefore UN-led processes are tantamount to "pseudo-multilateralism" (Krauthammer 1990). Donald Trump's "America First" motto provides the second explanation for why many Americans resist multilateral approaches and retreat into isolationism. Many ordinary Americans believe that the United States has failed to take care of its citizens at home and instead has devoted too many resources toward helping others abroad. As President Trump declared in his inaugural speech, "Every decision on trade, on taxes, on immigration, on foreign affairs will be made to benefit American workers and American families ... America will start winning again, winning like never before" (Trump 2017).

China's stance on international agreements is totally different because, in general, the Chinese government welcomes multilateral approaches to global challenges. The first reason for this acceptance is cultural in nature. In China, no individual person should act purely based on self-interest but instead must consider his or her responsibility to others, such as in the Confucian notion of the father who bears responsibility for his son and vice versa. In a group, responsibility for the group may require a sacrifice of an individual's own interests to help others if needed. Similarly, a country in a global context also bears responsibility for its own actions and must consider its impact on others. Second, to most Chinese citizens, international agreements enjoy the highest legitimacy because they are negotiated and adopted with the best interests of the entire world in mind. In China, the more people a policy or agreement benefits, the more support it will enjoy in the population. The third factor is China's sense of dignity in the world. China once was a large and powerful country internationally but perceives itself to have lagged behind others in recent decades. China strongly wishes to be respected by other countries again and desires to be seen as a responsible country in the global arena. China also welcomes the international climate agreements because their obligations reinforce China's domestic agenda, such as the need to shift to a more sustainable mode of economic development. Finally, leaders at the top of China's pyramidal and hierarchical structure have more power to make decisions about international agreements than do most leaders in other countries.

Revisiting the Myths

We can now understand why the myths mentioned in chapter 1 are misleading, or even completely wrong. Let us begin with prominent conspiracy theories. Future President Trump himself asserted on November 6, 2012, that "the concept of global warming was created by and for the Chinese in order to make U.S. manufacturing non-competitive" (Trump 2012). A citizen-businessman at the time, perhaps Trump had read the book published by US Senator James Inhofe earlier that year: *The Greatest Hoax: How the Global Warming Conspiracy Threatens Your Future*. Senator Inhofe, from the oil- and gas-producing state of Oklahoma, was the chairman or ranking member of the Senate Environment and Public Works Committee from 2003 to 2013.[2] In the book, Inhofe argues that the United States is an "over-regulation nation" that is being stifled by restrictive environmental regulation, drawing from his life experiences as a property developer and oil well driller. He writes, "It is clear that the global warming debate was never really about saving the world; it was about regulating the lives of every single American" (Inhofe 2012). Senator Inhofe was incensed that the Kyoto Protocol did not impose the same obligations on developing countries "like

China, India, Brazil and Mexico" as on the United States and declared that "Kyoto represented an attempt by certain elements in the international community to restrain U.S. interests" (ibid.).

There are conspiracy theories on the Chinese side as well. A popular Chinese talk show host, Larry Hsien Ping Lang, questioned the science of climate change in a televised speech in 2010 entitled "Climate Change Great Swindle"; the video clip earned nearly a million hits on China's YouTube equivalent. One comment from a viewer is telling: "These foreign bastards are so worried that China will rise and surpass the United States. Because they are jealous of China, they even made up lies about China and other developing countries . . . The scientists are all puppets controlled by politics. Copenhagen liars! American liars!" (Liu 2015). Reportedly, even some in Chinese government ministries concluded that the agreement pushed for in the Copenhagen negotiations was unfair to China, arguing that the Copenhagen Accord was "a conspiracy by developed nations to divide the camp of developing nations" (Dembicki 2017).

A pervasive myth in the West is that if China's leaders want to achieve a goal, they can just issue an order and it will be followed. An opinion article in the *Guardian*, for example, states that "China's top-down engineering-oriented approach means that it can set big goals and reach them" (Clifford 2015). In fact, as discussed throughout this book, ensuring local compliance with central government policies is one of the biggest challenges facing China's government leaders today. The central government does not directly enforce many policies because this is the primary function of the subnational levels of government in China, and the central government does not have the capacity to handle enforcement at the local level. A corollary to this Western myth is that if China's leaders do not issue an order, it is because they do not wish to do so. In other words, if the Chinese central government does not do something that a foreign government wishes it would do, it must be because the central government simply is refusing to do so. Instead, it clearly is possible that the central government is conscious of the fact that lower levels of government might not be able to achieve the objectives of such a policy, that certain bureaucracies or Politburo members are opposed to the policy, or, of course, that China's leaders simply do not think the policy is in China's best interest. In reality, there are many possible reasons that the central government may or may not choose to advance a new policy.

Somewhat the inverse of the above myth is the idea, sometimes advanced in China and elsewhere (even within the United States), that the United States is too democratic, with too many voices, and therefore has no ability to get anything done. Applied to the climate change issue, most Chinese cannot understand why the United States cannot pass a climate change law given that the United States is the second-largest emitter of GHGs in the world. In March 2016, China's official news organ released an op-ed by writer Zhu Junqing entitled "Trump's Rise Is the Fall of U.S. Democracy." The piece argued that the mere fact that Trump was the Republican contender illustrated "the malfunction of the self-claimed world standard of democracy" because the American people might vote not to "choose a president who is responsible to lead the country" but rather to "vent their grievance and anger over the reality" (Zhu 2016). This type of argument helps justify China's one-party rule, and also its mythology.

In fact, many other features of the US policymaking system contribute to the apparent inability to advance legislation in the United States, including legislation for climate change. Party politics have become so extreme that bipartisan consensus is now rare. The electoral system for the US Congress gives considerable weight to less populous, rural states in the US Senate because every state elects two senators regardless of population size. This structure is also true of the Electoral College, which technically elects the president of the United States and similarly gives greater weight to states with smaller populations. In addition, the constitutionally mandated separation of power across three branches of government means that each branch has the ability to limit the power of the other two branches, and they often work at cross-purposes to each other. However, as the climate change example demonstrates, determined presidents can advance policy using their executive authorities and US states are free to adopt more aggressive policies than the federal government's—and most have done so. Thirty-three states and territories of the United States have adopted renewable portfolio standards, for example, which the federal government does not have (DOE DSIRE 2016).

We conclude that there is much to admire and despise in the policy processes of both the United States and China. Looking toward the future, continued cooperation between the United States and China will be crucial to achieving sufficient emissions reductions to avoid the dangerous impacts of climate change because these two countries are the world's largest overall emitters of heat-trapping gases. Each country therefore must identify improved approaches to domestic climate policy, implement and enforce appropriate policies, and prove to the rest of the world that they have honored their commitments. If they can do so, the rest of the world will begin to trust that their own emission-reduction actions will be meaningful and effective, not undermined by noncompliance by the biggest emitters. China and the United States will also almost certainly be needed to lead the rest of the world through the arduous and time-consuming process of negotiation to achieve more ambitious global agreements over time. Indeed, the United States and China were crucial to the Paris Agreement process that culminated in 2015.

Conceptualizing American and Chinese Policymaking Behavior

The Chinese government exhibits strategic pragmatism today. Despite the overwhelming dominance of the CCP, communist ideology is more symbolic than motivational in China's policy process. The supreme goal for Chinese leaders is to maintain their power and the CCP's rule. They must deliver stability and an improved quality of life to maintain their legitimacy. The CCP promotes the idea of the China Dream, in which each Chinese citizen works together in harmony to restore China to the prosperity it enjoyed prior to the end of the Qing Dynasty. Chinese leaders therefore are relentlessly focused on the achievement of this long-term goal and will use authoritarian means to achieve it.

Although many aspects of the policy process remain fragmented (Lieberthal and Oksenberg 1988), China's economic power is consolidating, which allows the government to concentrate financial resources behind major priorities and to take a long view. China's government remains predominantly focused on domestic matters but is acutely aware of the broader global context. The Chinese government strategically harnesses the forces of globalization to serve its own interests (Gallagher 2014). The Chinese government works to develop and achieve strategic objectives through extensive use of planning. It can afford to think in five-year increments, decades, or even to mid-century because it expects to remain in power so long as it delivers gains to the Chinese people.

The term *strategic pragmatism* was first coined by Schmiegelow and Schmiegelow in 1989 to explain the performance of Japanese economic development. Some years later, Edgar Schein (1996) also employed the term to explain how economic development was promoted in Singapore through its Economic Development Board. To our knowledge, this term has not yet been applied in the Chinese context, although Zhao Suisheng's (2016) edited volume on Chinese foreign policy comes close. Zhao notes that pragmatist strategy is "ideologically agnostic, having nothing ... to do with either communist ideology or liberal ideas." Our use of the term is intended to describe China's approach to domestic policymaking more generally based on our case studies of climate change policy. Although the strategic pragmatist approach produces coherent policies that are developed through an elite consensus-building process, the public has little opportunity to influence Chinese policy, which may make it inherently unstable over the very long term.

In his book *Destined for War*, Graham Allison (2017) argues that "no one should be deluded into thinking that the [Chinese] regime today is a post-ideological movement solely concerned with its own power." Although each regime needs ideology to provide legitimacy to its rule, we argue that the CCP actually adopts a pragmatic attitude and approach to ideology. The CCP simply adjusts ideology to justify its actions, which means it relies more on nationalism than on communist theory, as it used to do. Chinese leaders are in avid pursuit of economic might. The CCP has long embraced a concept known as *national rejuvenation*, and uses it as an organizing principle for its legitimacy. The term can be traced to back to the Thirteenth National Congress report delivered by Zhao Ziyang (赵紫阳) in 1987 and it was subsequently followed by Jiang Zemin (江泽民) and Hu Jintao (胡锦涛). In Xi Jinping's period, however, the slogan was revived and has gained increasing importance in the context of rising nationalism. The CCP's rule in China is effectively an oligarchy and it acts like an "organizational emperor" (Zheng 2010). Xi Jinping's speech to the Nineteenth National Congress illustrates how the party organizes progress without rigid adherence to ideology: "Our Party united the people and led them in launching the great new revolution of reform and opening up, in *removing all ideological and institutional barriers* to our country and nation's development, and in embarking on the path of socialism with Chinese characteristics" (Xi 2017; emphasis added).[3]

Chinese climate policy is a clear manifestation of strategic pragmatism. The Chinese government leaders recognize, for example, that the clean energy industry is a means to strategic competitive advantage internationally in a form of green mercantilism (Gallagher 2014). Climate change and other forms of environmental pollution directly threaten the well-being of Chinese citizens because their health is affected by air and water pollution, and adequate supplies of fresh water are threatened by climate change. Sea-level rise, an increased frequency of storms, and stronger floods also threaten China's development. Acting responsibly to address climate change also earns China's leaders approbation from the international community. These are all reasons that the Chinese government acts progressively on climate change. These reasons ring true for the United States as well, but the lack of societal consensus about the need to act combined with the separation of power has resulted in slower progress.

The United States is characterized by a deliberative incrementalism. The democratic electoral system causes most leaders to have a near-term time horizon for their policy objectives. Each policy change is repeatedly contested in deliberative fashion through the legislative process and the interplay between the executive, legislative, and judicial branches, as envisioned in the US Constitution. Frequent shifts in party control of the executive and legislative branches lead to rapid swings in policies. Policies thus zigzag back and forth, forward and backward, before they eventually consolidate over many years into a norm supported by hard-won political consensus. Politicians wish to have immediate impact to show to their constituents, so incremental changes in policy are proudly touted to the public whether or not they are actually a significant step toward solving a real problem. Most elected leaders believe they cannot afford to think about longer-term strategic interests or the greater good, and for those members of the House of Representatives who are elected every two years, the time horizon is exceedingly short. Senators serve for six-year terms, and presidents for four-year terms (with a maximum of two terms).

The term *deliberative incrementalism* was coined by Jacobs (2013), who sought to explain why presidents' efforts to take arguments about why a policy is needed directly to the public can fail: opposing politicians are able to effectively counter them. Jacobs studies the notion first embraced by President Wilson that if ordinary citizens understand the need for a policy, they will demand it of their elected leaders in the Congress. Wilson pioneered the practice of having the president communicate directly with the public.

For our purposes, we do not emphasize the role of the president in particular, but rather the many voices that are expressed through all branches of government that cause advances in policymaking to slow down, and even temporarily reverse themselves. Contemporary presidents try to influence the public debate through their speeches, Twitter, and other mediums, but so do hundreds of Congress members and ordinary citizens. Lawsuits contesting policies that are filed by firms, individuals, or parts of the government are another form of deliberation that slows down the implementation of policy.

Although deliberative incrementalism sometimes produces policy incoherence in the short term, societal consensus sometimes emerges over time that results in a stable policy. Elimination of slavery and provisions for civil rights are two examples of policies achieved through emerging societal consensus over decades.

The last two American Presidents were elected on strongly idealistic grounds. For Trump, it was to "make America great again," a nostalgic, backward-looking dream. Barack Obama ran on the idea of "hope," which was completely intangible yet irresistibly alluring to many voters. Americans are congenital optimists and the American Dream still exists in American culture. This optimism about individual opportunity can mask measly progress in the near term. Alexis de Tocqueville wrote in *Democracy in America* ([1840] 2006) that Americans "all consider society as a body in a state of improvement." This optimism derives from American ambition, which Tocqueville believed came from individual equality because each American believes he or she is entitled to life, liberty, and the pursuit of happiness, as stated in the Declaration of Independence. Americans are optimistic about their futures because they believe each individual has the opportunity to advance despite the evident inequities in the structure of society.

American climate policy exhibits incrementalism through the hundreds of incremental regulatory steps that have been taken under the authority of the Clean Air Act and various energy policy acts at the federal level, and the thousands of laws enacted and regulatory steps taken at the state level. US energy efficiency standards, enacted since the 1970s, have caused millions of tons of carbon dioxide emissions to be avoided, and even though they were never intended to combat climate change alone, they are an example of how incremental steps can lead to long-term gains. The steady federal investments in energy innovation that contributed to the shale gas revolution are another example. The shale gas revolution enabled the US power sector to begin to shift from coal to gas, which also led to substantial reductions in CO_2 emissions in that sector.

Incrementalism is manifested in American climate policy through the repeated setting of medium-term targets that the government then fails to meet. Incrementalism is also evident in the leadership of individual US states and cities that show the federal government what is possible through their initiative, even though, due to their limited size, the impact is relatively small. Individual Americans make incremental steps by putting solar panels on their roofs and buying electric vehicles. Businesses and universities set voluntary goals and impose restrictions upon themselves to demonstrate that they can do well while still doing good.

What to Expect

For now, the United States will continue to behave erratically, as changes in different presidential administrations cause domestic and international climate policies to swing back and forth between progressive and regressive American political leadership. Consensus is steadily rising among the American public about the need to act on climate change, and there will likely be a tipping point that finally catalyzes Congress to pass long-term legislation, whether the catalyst is another major hurricane, widespread wildfires in the American West, or an exceptionally strong drought. If President Trump is defeated by almost any other politician, Republican or Democrat, the United States is likely to rejoin the Paris Agreement given the bipartisan chorus of voices that emerged in early 2017 to urge President Trump to remain in the deal. More progressive Republicans, many independents, and most Democrats will likely work to pass a national climate law as soon as Trump is no longer president.

China will continue to move steadily, inexorably, toward assuming a greater leadership role on climate change internationally. It will first try to dominate global markets in clean energy technologies to serve its domestic strategic industries. It will continue to aggressively implement a broad array of domestic climate change policies that are much more predictable and stable than their counterparts in the United States. China is likely to overachieve its goal to peak its emissions around 2030 and may actually bend its emission curve down within the next decade. China's sense of urgency for promulgating new domestic climate policies may be undermined by the reversals of the Trump presidency, but the directionality of China's climate policies is unlikely to change.

This book was intended to demystify the United States or China, whichever country each reader knows less about, and to advance mutual understanding about how and why policy outcomes occur. We particularly sought to explain how and why climate change policy is made given the urgency for each country's government to prevent massive disruption to the climate. It is our hope that over time our two countries can learn from each other's experiments and mistakes, and together blaze a new path toward true, sustainable prosperity.

Appendix: Chronology of Climate Policies in China and the United States

Major national climate policies in China since 2000

Sector	Policy	Dates	Type	Issuing ministries and policy number
Economy-wide				
	Specially Designated National Plan on Science and Technology Development in Tackling Climate Change during the 13th FYP	Updated in 2017, first issued in 2012 for the twelfth five-year plan period	Innovation	MOST, MEP, & CMA [2017] No. 120
	The plan is jointly released by MOST, MEP, and CMA.			
	The 13th Five Year Plan on Energy Development (2016–2020)	2016	Regulatory plan	No. 31 [2014] of the State Council; NDRC
	The latest plan provides an update on the targets set in the *Energy Development Strategy Action Plan (2014–2020)*. The new targets include a cap on annual primary energy consumption set at five billion tons of the standard coal equivalent by 2020, with a need to limit the annual growth rate of primary energy consumption to 2.5 percent. The annual coal consumption should be held below 4.1 billion tons until 2020. The share of nonfossil fuels in the total primary energy mix is to rise to more than 15 percent by 2020. The share of natural gas should reach 10 percent, while that of coal will be reduced below 58 percent. In addition, installed nuclear power capacity is to reach 58 GW by 2020, with an additional 30 GW expected to be under construction in 2020. Installed capacity of hydro, wind, and solar power in 2020 is expected to reach at least 340 GW (plus 40 GW pumped storage power), 210 GW (205 GW online, 5 GW offshore) and 110 GW (including more than 60 GW of distributed solar energy systems and 5 GW of thermal solar), respectively. Energy self-sufficiency should be above 80 percent. In addition, China aims to reduce carbon dioxide emissions per unit of GDP by 18 percent from 2015 levels by 2020.			

Plan	Year	Type	Reference
Energy Technology Revolutionary Innovation Action Plan (2016–2030) The objective of the plan is that by 2020, China should see a significant improvement in independent energy innovation, with major breakthroughs in key technologies and a decrease in foreign dependence for energy technology and equipment, key components, and materials. By the year 2030, a sound energy technology innovation system will be in place, with a capacity to support coordinated and sustainable development of China's energy industry. By then, China should be among the global powers in energy technology. The action plan also includes a *Roadmap of Key Innovation Actions for Energy Technology Revolution*, putting forward innovative objectives for 2020, 2030, and 2050, respectively.	2016	Innovation plan	NDRC Energy [2016] No. 53
China 13th Energy Technology Innovation Five Year Plan (2016–2020)	2016	Innovation plan	NEA Technology [2016] No. 397
Work Plan for the Pilot Construction of Climate Resilient Cities Jointly released by the NDRC and the MOHURD, the plan proposes to incorporate climate resilience indexes into the urban-rural planning system, construction plans and industrial development plans, build thirty climate-resilient pilot cities, improve average cities' climate-resilient management, and raise the proportion of green buildings to 50 percent by the year 2020.	2016	Plan	NDRC Climate [2016] No. 245
Work Plan for Greenhouse Gas Emission Control during the 13th Five-Year Plan Period (2016–2020) Aims to lower carbon dioxide emission per GDP unit by 18 percent of 2015 emission level by 2020.	2016, first issued in 2011 for the twelfth five-year plan period	Plan	No. 61 [2016] of the State Council

Major national climate policies in China since 2000 (continued)

Sector	Policy	Dates	Type	Issuing ministries and policy number
	Comprehensive Work Plan on Energy Conservation and Emission Reduction Mandatory energy-intensity-reduction targets were first allocated to local governments in 2007. The latest work plan published for the thirteenth FYP sets forth that by 2020, the national energy consumption per 10,000 RMB of GDP will be reduced by 15 percent compared with 2015; the total energy consumption will be capped at five billion tons of standard coal; and the total volatile organic compounds (VOC) emissions across the whole nation will be cut by more than 10 percent compared with 2015. The latest work plan also initiates the *100-1,000-10,000 Energy Conservation Program*, which aims to put the top one hundred energy-consuming enterprises in China under regulation of the central government, the top one thousand energy-consuming enterprises under the regulation of their respective provincial-level governments, and a further ten thousand plus high-energy-consuming enterprises under the regulation of lower-level governments.	2016, first published in 2007 and then updated in 2011	Regulatory plan	No. 74 [2016] of the State Council
	Energy Conservation Law	Revised in 2016, first issued in 1997	Law	State Council Presidential Order 48
	Administrative Measures for Energy Efficiency Labels China's national energy-efficiency-labeling system started with refrigerators and air conditioners. The revised measures specify the information to be included in the energy-efficiency labels and requests that manufacturers and importers use the labels on energy-consuming products listed in the corresponding catalog.	2016, first introduced in 2004	Informative, regulatory	NDRC & AQSIQ [2016] No. 35

Opinions of the CPC Central Committee and the State Council on Further Promoting the Development of Ecological Civilization	2015	Guideline	No. 12 [2015] of the CPC Central Committee
Overall Plan for the Structural Reform for Ecological Civilization The plan aims at gradually establishing the control system and the implementation mechanism of national total carbon emissions, establishing an effective mechanism to increase the forests, grasslands, wetlands, and ocean carbon sink, and strengthening international cooperation in response to climate change.	2015	Plan	No. 25 [2015] of the CPC Central Committee
National Plan on Climate Change (2014–2020) The plan sets a target of reducing carbon emissions per unit of GDP by 40–45 percent from 2005 level by 2020, increasing the percentage of nonfossil fuels in primary energy consumption to 15 percent and increasing the proportion of forest area and stock volume by 40 million hectares and 1.3 million cubic meters respectively from a 2005 baseline.	2014	Plan	NDRC Climate [2014] No. 2347
Notice on Organizing and Promoting Key Enterprises and Public Institutions to Report Greenhouse Gas Emissions	2014	Regulatory	NDRC Climate [2014] No. 63
National Strategy for Climate Adaptation Lays out clear guidelines and principles for climate change adaptation and proposes some specific adaptation goals.	2013	Plan	NDRC Climate [2013] No. 2252

Major national climate policies in China since 2000 (continued)

Sector	Policy	Dates	Type	Issuing ministries and policy number
	HCFC Phase-out Management Plan (HPMP) Since 2011, China has been implementing the first stage of its HPMP in industrial and commercial refrigeration, targeting at phasing out 3,386 ODP tons of HCFCs by 2015. The Chinese government also finalized proposal for the Stage II HPMP in 2016, with a focus on natural refrigerant technologies. Stage II proposes a phase-out of 4,749 ODP tons of HCFCs by 2020 and an additional 4,684 ODP tons by 2026 to assist the Government of China in meeting 35 percent and 67.5 percent reduction targets by 2020 and 2025, respectively. The MEP issued its *Circular on Strict Management of HCFC Production, Sales and Consumption* in 2013, requiring quota permits from all enterprises producing HCFCs and consuming over one hundred metric tons (mt) of HCFCs and registration at local Environmental Protection Bureaus for enterprises consuming less than 100 mt.	2011	Regulatory	MEP [2013] No. 179
	Circular Economy Promotion Law	2008	Law	State Council Presidential Order 4
	National Climate Change Programme The program, China's first global warming policy initiative, outlines objectives, basic principles, key areas of actions, and policies and measures to address climate change for the period up to 2010.	2007	Guideline	No. 17 [2007] of the State Council

Promotion of Circular Economy In 2005, the State Council issued **Suggestions on Accelerating the Development of Circular Economy**, which was China's first document that supported promotion of circular economy implementation from a national level. In 2010, the NDRC issued the *Guidelines for Making Plans for Circular Economy Development*, suggesting that local governments should develop the circular economy according to their specific circumstances. In 2013, the State Council issued *Development Strategy and Immediate Action Plan of Circular Economy*, setting goals for China's circular economy development in different stages. The NDRC also issued *Plan for the Promotion of Circular Economy* documents in 2014 and 2015, which include actions and targets to use resources (water, metals, land and coal) more efficiently and to better manage resources and waste in industries, agriculture and cities. In addition, MIIT has released six editions of *Catalogue of Remanufactured Products* to promote the use of remanufactured products.	2005	Guideline	No. 22 [2005] of the State Council, NDRC [2010] No. 311, No. 5 [2013] of the State Council, NDRC Environment and Resources [2015] No. 769, MIIT [2016] No. 67
Medium- and Long-Term Energy Conservation Plan The plan includes energy conservation targets up to 2020. Energy consumption per RMB 10,000 GDP is expected to drop to 1.54 tons of coal equivalent in 2020, with an annual average energy conservation rate of 3 percent from 2003 to 2020. And by 2020, energy consumption per unit of major product production is expected to reach or approach the international advanced level. It also identifies key fields and key projects for improving energy efficiency.	2004	Plan	NDRC Environment and Resources [2004] No. 2505

Major national climate policies in China since 2000 (continued)

Sector	Policy	Dates	Type	Issuing ministries and policy number
Transportation				
	The 13th Five-year Plan for Energy Conservation and Emissions Reduction in Road and Water Transport The plan provides specific energy-consumption and CO_2-emissions-reduction goals for different modes of transport for 2020: For operating vehicles: Energy consumption per unit of transport volume should fall 10 percent by 2015 and 16 percent by 2020 from 2005 levels; CO_2 emissions per unit of transport volume should fall 11 percent by 2015 and 18 percent by 2020 from 2005 levels. For operating ships: Energy consumption per unit of transport volume should fall 15 percent by 2015 and 20 percent by 2020 from 2005 levels; CO_2 emissions per unit of transport volume should fall 16 percent by 2015 and 22 percent by 2020. For urban transport per person: Energy consumption should fall 18 percent by 2015 and 26 percent by 2020 from 2005 levels; CO_2 emissions intensity should fall by 20 percent by 2015 and 30 percent by 2020 from 2005 levels. Prior to this, China also issued *Medium- and Long-Term Outline about Energy Conservation in the Fields of Road and Water Transportation* in 2008, *Guiding Opinions on Establishing a Low-Carbon Transport System,* and *12th Five-year Plan for Energy Conservation and Emissions Reduction in Road and Water Transportation* in 2011.	2016	Plan	MOT [2016] No. 94

R&D Program of New-Energy Vehicles (2016–2020)
The Electric Vehicle Major Project has been an important component of the 863 Program since the tenth five-year plan period. In 2015, MOST solicited opinions for a plan to support the research and development of new energy vehicles as a key major project over the next five years. According to the plan, in 2020, the battery module should have an energy intensity of 300 Watt-hours/kilogram or more, and fuel cell vehicles should achieve a market scale in the "thousands" by 2020.

2016, first introduced in 2001

Innovation

MOST [2016] No. 305

Adjustment to Subsidies for New Energy Vehicles
The policy adjustment was jointly issued by MOF, MOST, MIIT, and the NDRC in December 2016. It supplements *Financial Support Policy for New-energy Vehicles, 2016–2020* issued in 2015 and represents the sixth adjustment to the original policy introduced in 2009. It requests that subsidies for pure electric vehicles and plug-in hybrid vehicles be reduced by 20 percent in 2017 to 2018 from 2016 levels, and 40 percent in 2019 to 2020 from 2016 levels. Following this adjustment, the subsidies from the central government range from RMB 20,000–44,000 per vehicle for pure electric and hybrid electric vehicles. Subsidy standards for fuel cell electric vehicles remain unchanged, ranging from RMB 200,000 to 500,000 per car. In September 2013, MOF announced a long-anticipated renewal of China's electric vehicle subsidies. Consumers who purchase EVs can get up to RMB 60,000 (USD 9,400) for pure electric cars with a range over 250 km, and RMB 50,000 and 35,000 for EVs with ranges over 150 km and 80 km, respectively.

2016, first introduced in 2009 and then updated in 2013 and 2015

Fiscal

MOF [2016] No. 958

Major national climate policies in China since 2000 (continued)

Sector	Policy	Dates	Type	Issuing ministries and policy number
	Emission Standards for New Passenger Cars and Light-Duty Commercial Vehicles In December 2016, MEP released the final rule of the Stage 6 Limits and Measurement Methods for Emissions from Light-Duty Vehicles. The China 6 standard, which is to take effect beginning on July 1, 2020, is one of the most stringent emission standards around the world for the post-2020 time frame. Unlike the previous standard phases, which closely follow the European emission standards, the China 6 standard combines best practices from both European and US regulatory requirements (California regulations) in addition to creating its own.	2016, first introduced in 1999	Regulatory	MEP [2016] No. 79
	Tax-Exemption Policy for New-Energy Vehicles In 2015, a new tax-exemption policy was jointly issued by MOF, MIIT, and the State Administration of Taxation. It is an updated version of the policy introduced in 2012. According to this policy, new-energy vehicles and ships will be exempted from vehicle and vessel taxes.	2015, first introduced in 2012	Fiscal	MOF, SAT, & MIIT [2015] No. 51

| Demonstration and Promotion of Energy Efficient and Alternative Energy Vehicles
This was first known as the Ten Cities, Thousand Vehicles program, which was intended to stimulate electric vehicle development through large-scale pilots in ten cities with government subsidies, focusing on deployment of electric vehicles for public fleet applications. The program has since been expanded to eighty-eight cities. *Opinions on Accelerating the Promotion of the Application of New Energy Vehicles in the Transportation Industry* released by MOT in 2015 set a target of three hundred thousand new-energy vehicles on China's roads by 2020: two hundred thousand new-energy buses and one hundred thousand new-energy taxis and delivery vehicles. | 2015, first introduced in 2009 and then updated in 2013 and 2014. | Fiscal | MOF & MOST [2009] No. 6, MOT [2015] No. 34 |

Major national climate policies in China since 2000 (continued)

Sector	Policy	Dates	Type	Issuing ministries and policy number
	Vehicle Fuel Economy Standards China's first-ever fuel-consumption standards for passenger vehicles were adopted in 2004. The latest Phase 4 Passenger Car Fuel Consumption Standard released by MIIT in 2014 regulates domestically manufactured and imported new passenger cars sold in China from 2016 to 2020. It projects an overall fleet-average fuel consumption of 5 L per 100 km for new passenger cars in 2020. The Phase IV regulation includes both vehicle-maximum fuel-consumption limits and a corporate-average fuel-consumption (CAFC) standard for each manufacturer based on vehicle curb weight distribution across the manufacturer's fleet. Manufacturers and importers must meet both standards. China introduced fuel-consumption standards for light-duty commercial vehicles in 2007 and last updated the standards in 2015. The latest standards, which are to take effect on January 1, 2018, are 18–27 percent more stringent than the 2007 standards and are expected to reduce the fuel-consumption level of new light-duty commercial vehicles by 20 percent in 2020 as compared to 2012. MIIT first announced its plan to develop fuel-consumption standards for commercial heavy-duty vehicles in 2008. The Phase 2 heavy-duty vehicle fuel consumption standards were finalized in 2014. By July 1, 2015, all new commercial HDVs sold in China (except specialized vocational vehicles) were required to comply with the Phase 2 standards.	2015, first introduced in 2004 and amended in 2014	Regulatory	GB 20997–2015

Energy-Saving and New-Energy Vehicles Industry Development Plan (2012–2020)
The plan targets the production of five hundred thousand BEVs and PHEVs by 2015, with the production capacity to grow to two million units and the cumulative production and sales of more than five million of those types by 2020. To ensure the enforcement of the plan, the State Council issued "Guidance on Accelerating the Popularization and Application of New Energy Vehicles" in 2014.

2014, first released in 2012 | Plan | No. 35 [2014] of the State Council

Power

Emissions trading system (ETS)
Pursuant to *National Plan on Addressing Climate Change (2014–2020)*, a national carbon-emission-trading market will be formed to lower the cost of achieving GHG reduction goals.

Scheduled launch in 2017 | Market | Under development

Implementation Plan for the Licensing System to Control Pollutant Emission
The plan requires all stationary sources of pollution in China to be licensed by 2020 to further curb emissions. All companies should apply for the license before undertaking industrial production, allowing the authorities to monitor pollution in advance. The discharging policy gives companies a pollutant-discharge permit, which covers specifics such as the variety of pollutants, concentration, and amounts allowed. Those that violate the restrictions will face strict penalties ranging from suspension of operations to criminal charges. The policy is scheduled to come into force by the end of 2016 in thermal power plants and papermaking companies, and then expand to cover fifteen major industries that discharge air and water pollutants by 2017.

2016 | Regulatory | No. 81 [2016] of the State Council

Major national climate policies in China since 2000 (continued)

Sector	Policy	Dates	Type	Issuing ministries and policy number
	Administrative Measures on Protective Full Purchase of Renewable Energy Generation This document mandates that grid companies purchase output from renewable generators, at least up to an allocated number of hours. The NDRC and the NEA will be responsible for planning annual allocations of operational hours for each type of renewable generation in regions of the country that have been experiencing curtailment. Based on different circumstances, it calls on conventional power generators or the grid companies to compensate renewable energy generators for curtailment.	2016	Regulatory	NDRC Energy [2016] No. 625
	Guiding Opinions on Promoting Electric Energy Substitution This guideline states that electric power should replace some 130 million tons of dispersed coal and fuel between 2016 and 2020, which should drive the electricity-generation-to-coal consumption rate up by 1.9 percent and the electric-energy-to-terminal-energy consumption rate up by 1.5 percent to 27 percent.	2016	Guideline	NDRC Energy [2016] No. 1054
	Guiding Opinions on the Establishment of a Target Setting System for the Development and Utilization of Renewable Energy For the first time, portfolio standards for nonhydropower renewable energy were issued for provinces and municipalities (for the year of 2020).	2016	Regulatory	NEA [2016] No. 54

Notice on Solar PV Deployment Management and Introduction of Competitive Bidding Starting from January 1, 2016, the bidder with lowest prices (and other indicators) will be awarded the right to build a PV power plant.	2016	Fiscal	NEA New Energy [2016] No. 14
The 13th Five Year Plan for Power Sector Development It is estimated that by 2020, Chinese electric power consumption will reach 6,800 TWh of electricity, increasing on average by 3.6–4.8 percent each year. The per capita use is expected to reach approximately 5,000 kWh by 2020. The plan outlines specific targets for power generation from each type of energy resource, as well standards for system upgrades and reforms.	2016, Last plan for the power sector was released in 2001	Plan	Document number unknown Detailed goals can be found at: http://en.cnesa.org/latest-news/2016/11/22/power-sector-reforms-announced-in-chinas-13th-five-year-Plan Because most details of this plan were covered in the thirteenth FYP for energy development, this document may be seen as a component of that overall guideline.

Major national climate policies in China since 2000 (continued)

Sector	Policy	Dates	Type	Issuing ministries and policy number
	The 13th Five-Year Plan on the Development of Renewable Energy Provides guidelines for the development of various renewable energies, including solar, wind, hydropower, biomass, and geothermal energy. The plan projects the investment for renewable energy to reach the amount of RMB 2.5 trillion (approximately USD 380 billion) for the thirteenth five-year period, and the annual usage of renewable energy will be 730 million tons of coal equivalent. The plan also calls for the establishment of a nationwide mechanism for trading Renewable Energy Green Certificates (or simply Green Certificates), which will be used to document a power-generation enterprise's use of nonhydropower renewable energy. In 2020, the electricity generated by nonhydropower renewable energy is projected to account for at least 9 percent of all electricity generated by each power-generation enterprise. In parallel to this overall plan, individual thirteenth Five-year plans for hydropower, wind, bioenergy, solar, ocean energy and renewable energy development were also released.	2016; the last mid- to long-term program of renewable energy development and development plan for specific types of renewable energy were issued in 2007; development plans for overall and specific types of renewable energy were also issued for the twelfth FYP in 2012	Plan industrial	NDRC Energy [2016] No. 2169

	First policy issued in 1999, repeatedly reiterated, most recently in 2016	Regulatory guidelines	MEP [2015] No. 164, NDRC Energy [2016] No. 565, NEA Power [2016] No. 244, NEA Power [2016] No. 275, NDRC Energy [2016] No. 855, NDRC Energy [2016] No. 617
Promoting clean and efficient development of coal-fired power generation Recent policies in this regard include the following: *Full Implementation of Ultra-Low Emission and Energy Saving Transformation of Coal-fired Power Plant*, *Notice on Promoting of an Orderly Development of China's Coal-fired Power Plants* (March 2016), *Notice on Further Regulation of Coal-Fired Power Planning and Construction*, *Notice on Canceling a Number of Coal Power Projects Do Not Meet Approval Conditions*, *Notice on Further Eliminating Backward Production Capacity of Coal-fired Power Industry* and *Regulations of Combined Heat and Power Generation*, and others. According to these guidelines, upgrading of coal-fired power plants to achieve ultra-low emissions and energy conservation should be completed in 2017 in the eastern region, in 2018 in the central region, and by 2020 in the western region. A halt to construction of coal-fired plants in thirteen provinces in which capacity is already in surplus is ordered, including major coal producers such as Inner Mongolia, Shanxi, and Shaanxi. A further fifteen provinces are required to delay construction of already approved plants. In provinces with an electricity gap, priority should be given to the development of local nonfossil energy-generation projects, with the intent to use transprovincial energy transfers and other demand-side management approaches that could curtail the demand for new coal-fired generating plants. Thermal power generators that have gone through many years of service and are not energy-efficient, safe, or environmentally sound should be phased out, and condensing units below 300 MW which have operated for at least twenty years, as well as condensation extractors for thermal power plants that have operated for twenty-five years or more, should be shut down. Altogether, 20 GW of backward coal-fired power units are expected to be eliminated during the thirteenth FYP.			

Major national climate policies in China since 2000 (continued)

Sector	Policy	Dates	Type	Issuing ministries and policy number
	Feed-in Tariff for Renewable Energy The latest update on the FIT for renewable energy is *Notice on Adjustments to Feed-in Tariffs for Onshore Wind and PV Power*, released by the NDRC in December 2016. According to this notice, the 2017 benchmark feed-in tariff for PV is between RMB 0.65/kWh (USD 0.094/kWh) and RMB 0.85/kWh, depending on region—representing a cut of between 13 and 19 percent from 2016 levels. The 2018 benchmark feed-in tariff for onshore wind power will range from RMB 0.40/kWh and RMB 0.57/kWh, representing a 15 percent cut from 2016 levels. The FIT for new distributed PV is unchanged in 2017 at 0.42 RMB/kWh, as is the offshore wind rate: 0.85 RMB/kWh for offshore wind power projects and 0.75 RMB/kWh for intertidal wind power projects. According to the notice, the latest step-down in support reflects continuing reductions in deployment costs for solar and wind plants.	Updated in 2016, introduced for wind power in 2003 and updated regularly; introduced for two solar PV power plants in 2008, with rates updated regularly	Fiscal	NDRC directive [2016] No. 2729
	Several Opinions on Further Deepening the Reform of the Electric Power System The reform plan seeks to encourage competition in the power sector and calls for a revamp of the existing pricing system. The plan allows gradual infusion of social capital in the power sales and newly added distribution business, while the electricity transmission business will remain with power grid companies. Foreign capital infusion is allowed in all the fields that are not on the negative list, and the same also does not need approval from the government.	2015	Guideline	No. 9 [2015] of the CPC Central Committee

National Solar Subsidy Program
In March 2009, China announced its first solar subsidy program, the building-integrated photovoltaics (BIPV) subsidy program, offering RMB 20/watt for BIPV systems and RMB 15/watt for rooftop systems upfront. In July 2009, the Golden Sun Demonstration Project, the second national solar subsidy program, was launched by MOF, MOST, and the NEA. The project was to provide upfront subsidies for qualified demonstrative PV projects from 2009 to 2012.

Fiscal — 2009 — MOF [2009] No. 129, MOF, MOST, & NEA [2009] No. 397

Renewable Energy Law
The original law, which took effect in January 2006, was aimed at "optimizing the country's energy structure and safeguarding energy security." It covered subsidies, pricing management, and supervision measures. The revised Renewable Energy Law launches a "protective full-amount acquisition system." Although the 2005 law contains similar requirements for state power grid enterprises to buy the total amount of power produced by renewable energy sources, it is said to be lacking in detail and therefore difficult to implement. Electricity grid enterprises are required to reach agreements with renewable energy power-generation enterprises that have obtained administrative permits or made a filing with the government to purchase all the renewable energy power they produce that satisfies the technical standards for grid synchronization. Power enterprises refusing to buy power produced by renewable energy generators will be fined up to an amount double that of the economic loss of the renewable energy company.

Industrial / Fiscal / Regulatory — Amended in 2009, first issued in 2005 — National People's Congress, http://english.gov.cn/archive/lawsregulations/

Major national climate policies in China since 2000 (continued)

Sector	Policy	Dates	Type	Issuing ministries and policy number
Industrial				
	Program for the Construction of an Energy-Saving Standard System NDRC, Administration of Quality Supervision, Inspection and Quarantine (AQSIQ), and the Standardization Administration launched the *One Hundred Energy Efficiency Standard Promotion Program* in 2012. As of January 2017, a total of 104 compulsory energy-consumption standards and seventy-three mandatory energy-efficiency standards have been published. This new program aims at covering all major energy intensive industries and products and enabling 80 percent of China's energy-efficiency standards to be on par with international standards by 2020.	2017	Regulatory	NDRC Environment and Resources [2017] No. 83
	The 13th Five-Year Plan for Shale Gas (2016–2020) The plan sets the target of tapping shale gas resources trapped in reservoirs up to 3,500 meters underground during these five years and producing thirty billion cubic meters of gas by 2020. The plan also calls for further increasing production to between eighty and one hundred billion cubic meters by 2030.	2016	Plan	NEA Oil & Gas [2016] No. 255
	The 13th FYP for Developing Energy Saving and Environmental Protection Industries By 2020, the added value of the energy conservation and environmental protection industries is to account for 3 percent of gross domestic product, becoming one of the pillar industries for the domestic economy.	2016, last FYP published in 2012	Guideline	No. 19 [2012] of the State Council, No. 30 [2013] of the State Council

Policy	Year	Type	Reference
Shale Gas Industrial Policy In 2016, the NEA approved *Development Plan for Shale Gas (2016–2020)*. The NEA released China's first shale gas industrial policy in 2013, pursuant to China's *12th Five-Year (2011–2015) Plan for Shale Gas*. The policy affirmed that shale gas fell within China's strategic emerging industries and called for tax-incentive policies for the shale gas industry.	2016, first issued in 2013	Guideline	NEA [2013] No. 5, NEA Oil and Gas [2016] No. 255
Industrial Green Development In 2016, MIIT released *Industrial Green Development Plan (2016–2020)* to promote green manufacturing through green supply chain and support the fulfillment of goals set in the thirteenth FYP and *Made in China 2025*. Prior to this, the *Special Action Plan on Green Industrial Development* was issued every year from 2012 to 2015 to promote the transformation and upgrade of traditional industries. Related to this, the *12th Five-Year Plan on Industrial Energy Conservation* was issued in 2012.	2016, first issued in 2012	Plan	MIIT [2016] No. 225
Measures for the Subsidy Funds for Energy Conservation and Emission Reduction	2015	Fiscal	MOF [2015] No. 161
Notice on Issuing Subsidies for Exploring and Utilizing Shale Gas In 2015, MOF announced that a subsidy of RMB 0.3 (USD 0.049) would be offered for every cubic meter of shale gas developed by enterprises during the 2016–2018 period, down from RMB 0.4 provided for the 2012–2015 period. From 2019 to 2020, the subsidy will be further lowered to RMB 0.2 for every cubic meter of shale gas exploration.	Renewed in 2015, first introduced in 2012	Fiscal	MOF [2015] No. 112

Major national climate policies in China since 2000 (continued)

Sector	Policy	Dates	Type	Issuing ministries and policy number
	Industrial Transformation and Upgrading In 2011, the State Council disseminated *Plan for Industrial Transformation and Upgrading (2011–2015)* to promote green and low-carbon industrial development. In 2015, China sped up the optimization of the industrial structure with the introduction of *Made in China 2025*, which sets forth such strategic tasks as improving innovative design capability, enhancing energy efficiency, promoting green transformation, and upgrading and resolving overcapacity in traditional industries.	2015, previous plan issued in 2011	Plan	No. 47 [2011] of the State Council, No. 28 [2015] of the State Council
	Action Plan for Clean and Efficient Use of Coal (2015–2020) According to the plan, China will raise its raw coal selective ratio to be above 70 percent by 2017 and 80 percent by 2020. It also plans to cut the average coal consumption of existing coal-fired power-generation units to below 310 g/kwh by 2020. By then, power coal use is expected to make up more than 60 percent of total coal use. It also requires major coal-consuming sectors to improve technologies for efficient and clean use of coal and accelerate elimination of outdated furnaces and boilers.	2015	Industrial plan	NEA [2015] No. 141
	Regulation for Energy-Saving and Low-Carbon Products Certification A tentative regulation was issued by the NDRC in 2013 before the joint release of this final regulation by AQSIQ and the NDRC. By March 2016, China had published two catalogs of certified products.	2015	Regulatory	GAQSIQ Decree No. 168

Management Measures for Certification of Energy-Saving and Low-Carbon Products	Updated in 2015, first introduced in 1999	Regulatory	GAQSIQ Decree No. 168
Action Plan for Retrofitting and Upgrading Coal-Fired Power Plants (2014–2020) The plan strengthens the energy-efficiency and pollutants-emission standards applied to coal power plants. Coal power plants with capacity over 600 MW are required to achieve the efficiency target of 300 g of coal equivalent/kWh by 2020. The coal power plants under construction or planned are required to reach the same level of pollutants emission as natural gas power plants.	2014	Regulatory	MEP [2015] No. 164
Notice of Publishing the Implementation Plan for the Energy Efficiency Leader Scheme The program aims to set up a long-term mechanism to incentivize energy-efficient "leaders" and to increase the level of energy-efficiency among high-energy-consuming products and equipment, high-energy-consuming industries, and public institutions. The scheme will raise current standards of energy-efficiency through incentive programs and industry benchmarks.	2014	Informative	NDRC Environment and Resources [2014] No. 3001
Resource Tax Reform China reformed its resource tax on crude oil and natural gas in 2011, and set a new resource tax rate on coal in 2014. Following these reforms, China began to levy a resource tax on crude oil, natural gas, and coal based on the retail price rather than production to promote the more efficient use of resources. The rates were set to be between 5 and 10 percent for crude oil and natural gas, and between 2 and 10 percent for coal.	2014 and 2011	Regulatory Fiscal	MOF [2011] No. 114, MOF [2014] No. 72

Major national climate policies in China since 2000 (continued)

Sector	Policy	Dates	Type	Issuing ministries and policy number
	Airborne Pollution Prevention and Control Action Plan (2013–2017) The plan proposes to improve overall air quality across the nation through five years, reduce heavy pollution by a large margin, and make obvious improvement of air quality in Beijing-Tianjin-Hebei Province, the Yangtze River Delta, and the Pearl River Delta. By 2017, the level of inhalable particulate matter in cities above the prefecture level is to be dropped by at least 10 percent from the 2012 level and the days with good air quality will be increased year on year. The level of fine particulate matter in Beijing-Tianjin-Hebei Province, the Yangtze River Delta, and the Pearl River Delta will be cut by 25, 20, and 15 percent, respectively, and the annual concentration of fine particulate matter in Beijing will be kept at 60 mcg /m^3. Coal consumption as a proportion of the entire energy mix should decline from the current 68 percent to 65 percent in 2017.	2013	Regulatory	No. 37 [2013] of the State Council
	Guideline Catalog for Industrial Restructuring This guideline aims to achieve China's target for conserving energy and reducing emissions by optimizing and upgrading its industrial structure.	Updated in 2013, first issued in 2005	Informative	NDRC directive [2013] No. 21

Capacity Elimination Program The level of capacity cuts for most industries rose substantially in 2010, along with the issuance of *Notice of the State Council on Further Strengthening the Elimination of Backward Production*. In 2012, MIIT issued *Notice of Issuing the Objectives of Eliminating Backward Production Capacity in 19 Industrial Sectors*. The five industries with severe excess capacity identified in 2012 include steel, cement, aluminum, flat glass, and shipbuilding. In 2013 and 2014, the first and second lists were issued of enterprises to be eliminated in the nineteen industrial sectors. In 2013, the State Council issued *Guidelines to Tackle Serious Production Overcapacity*, laying out the measures to deal with the problem. During the thirteenth FYP, China plans to phase out eight hundred million tons of backward coal capacity per year while adding 0.5 billion tons per year in advanced production capacity. Meeting its 2016 target ahead of schedule, China plans to reduce steel output by an additional one hundred million tons to 150 million tons by 2020.	2013, national level order first issued in 1999 and became a staple of government proclamations in 2007, with specific lists of affected factories and categories being updated regularly	Regulatory	No. 7 [2010] of the State Council, MIIT directive [2012] No. 159, No. 41 [2013] of the State Council
Promotion of Energy Efficient Products and Technologies Since 2012, MIIT has published *Energy Efficiency Star Certified Product Catalogue* every year. Specific energy-efficient product or technology catalogs have been issued for electromechanical devices, electronic motors, industrial boilers, internal-combustion engines, energy-saving technologies in telecommunication, and other low-carbon technologies.	2012	Informative Voluntary	MIIT [2016] No. 59

Major national climate policies in China since 2000 (continued)

Sector	Policy	Dates	Type	Issuing ministries and policy number
	Top-1,000 (then Top-10,000) Energy-Consuming Enterprises Program China's mandatory energy-conservation target-setting policy for large energy users, known as the Top-10,000 program, was introduced in 2011 under the twelfth FYP as an expansion of its successful predecessor, the Top 1,000 program, which operated between 2006 and 2010. The Top-10,000 program expanded the number of enterprises and covered two-thirds of China's total energy consumption. It achieved the energy-saving target of 250 million tons of coal equivalent one year before the 2015 deadline.	2011, first introduced in 2006	Regulatory	NDRC Environment and Resources [2006] No. 571 NDRC Environment and Resources [2011] 2873
	Energy-Saving Products Benefiting the Public The project provided a promotional list of energy-efficient products that covered three major categories—household appliances, vehicles, and industrial products—with fifteen varieties and about one hundred thousand types of energy-efficient products. The central government allocated more than RMB 40 billion for the implementation of the project.	2009, concluded in 2013	Fiscal Regulatory	MOF & NDRC [2009] No. 213

	Year	Type	Document No.
Differential Electricity Price Policy The policy requires a punitive electricity price (a surcharge of RMB 0.2/kWh) on eight energy-intensive industries. In 2010, the NDRC further increased punitive electricity prices on energy-intensive industrial enterprises. The electricity consumed by outdated enterprises will be charged a surcharge of RMB 0.1/kWh and electricity consumed by restricted enterprises will be charged an additional RMB 0.3/kWh.	2006	Regulatory	No. 77 [2006] of the General Office of State Council
Green finance			
Guiding Opinions on Building a Green Financial System The guidelines include a series of policy incentives to support and incentivize green investment. These incentives include relending operations by the People's Bank of China, specialized green guarantee programs, interest subsides for green loan-supported projects, support for introduction of the PPP model in the green industry, and the launch of a national-level green development fund. The guidelines emphasize the importance of building and improving the unified rules and regulations for green bonds and taking measures to reduce the financing cost of green bonds. The guidelines also emphasize the development of green insurance and require a further expansion in international cooperation on green finance.	2016	Guideline	No. 228 [2016] of the People's Bank of China

Major national climate policies in China since 2000 (continued)

Sector	Policy	Dates	Type	Issuing ministries and policy number
	Preferential Tax Policies for Renewable Energy In 2001, value-added tax (VAT) for wind power cut in half, to 8.5 percent (normal rate: 17 percent). In the same year, a circular determined that VAT collected for using municipal solid waste for power generation would be refunded back to the producer. In 2003, the VAT for biogas production was also reduced, to 13 percent. Since 2005, the VAT levied on small-scale hydropower plants has been 6 percent of income, and the VAT collected from some large hydropower plants has been returned to companies. Fuel ethanol produced by certified enterprises has been exempted from consumption tax and VAT. The import duties and import VAT have been gradually exempted for renewable energy equipment imported by domestic and later foreign-invested renewable energy projects. Since the enactment of China's new corporate income tax in 2008, tax breaks have been granted to companies with synergistic utilization of resources or investment in technology for environmental protection, energy, and water conservation. In 2013, MOF announced that 50 percent of the VAT refund was to go to producers of solar PV power. The policy was extended in 2016. Newly added wind farms after 2015 are eligible for a 50 percent refund on their VAT.	2016, first introduced in 2001, with new policies being added over the years	Fiscal	MOF & SAT [2016] No. 81

	Year	Type	Reference
Energy-Efficiency Credit Guidelines The guidelines put forward feasible instructions in terms of the characteristics of energy-efficiency projects, the priorities in energy-efficiency credit businesses, business entry, key issues in risk examination, process management, product innovation, and so on.	2015	Guideline	CBRC & NDRC [2015] No. 2
Green Financial Bond Directive The directive and the associated Green Bond–Endorsed Project Catalogue set out the official requirements for what projects qualify as green, management of proceeds, and reporting.	2015	Guideline	No. 39 [2015] of the People's Bank of China
Interim Measures on the Management of the Additional Renewable Energy Surcharge Fund These measures set the rates of subsidies for the operation and maintenance costs to be incurred by connecting renewable energy to the grid, depending on the distance of power transmission: 0.01 RMB/kWh for less 50 km, 0.02 RMB/kWh for 50–100 km, and 0.03 RMB/kWh for 100 km or longer.	2012	Fiscal	MOF [2012] No. 102
Green Credit Guidelines The guidelines encourage Chinese banks to lend more to energy-efficient and environmentally sustainable companies and less to polluting and high-energy-consuming enterprises. The guidelines require banks to measure and control environmental and social risks in lending and will be applied to all lending—both domestic and overseas.	2012	Guideline	CBRC [2012] No. 4

Major national climate policies in China since 2000 (continued)

Sector	Policy	Dates	Type	Issuing ministries and policy number
	Interim Measures for the Administration of the Collection and Use of the Renewable Energy Development Fund The renewable energy development fund shall include the special-purpose fund appropriated by the public budget of the national finance and the income from surcharges on renewable energy power prices as collected from power users according to law.	2011	Fiscal	MOF [2011] No. 115
	Provisional Measures for Administration of the Financial Reward Capital for the Innovation of Energy-Saving Technology The prerequisite for getting the reward is that companies must have consumed more than twenty thousand tons of coal equivalent (TCE) per year before adopting energy-efficient technologies and save 5,000 TCE after their tech transformation. Companies in eastern regions will be rewarded with RMB 240 (USD 36.92) per TCE saved, whereas those in central and western regions will be given RMB 300 per TCE saved.	Revised in 2011, first issued in 2007	Fiscal	MOF & NDRC [2011] No. 367

Land use

	Year	Type	Document No.
Plan on National Forest Management (2016–2050) The plan sets the objective of raising the forest cover to over 26 percent by 2050 and increasing the forest stock volume to 23 billion cubic meters.	2016	Plan	SFA [2016] No. 88
The 13th Five Year Plan for China's National Forestry Development (2016–2020) The plan envisages China's national forest stock will reach 1.4 billion cubic meters, the forest cover will rise to just over 23 percent, and the total value of forestry industry output will amount to RMB 8.7 trillion.	2016	Plan	SFA [2016] No. 22
Instruction Opinions on Advancing the Forestry Carbon Sink Trade	2014	Guideline	SFA [2014] No. 55
Outlines of National Forestation Plan (2011–2020) The plan sets the targets of increasing forest coverage to over 23 percent, forest stock volume to over 15 billion cubic meters, forestry output to 10 trillion RMB, and the compulsory tree planting rate to 70 percent by 2020.	2011	Plan	NAC & SFA [2011] No. 6
The Program of Constructing a National Monitoring System on Forest Sinks SFA issued the *Interim Management Measures for Quantifying and Monitoring Forest Carbon Sinks* in 2010 and started to build the forest inventory system in seventeen pilot provinces/municipalities. By the end of 2015, the program had covered twenty-five provinces, autonomous regions, and municipalities, Xinjiang Production and Construction Corps, and the country's four biggest forest industry groups. A basic database of forestry carbon sinks was built up.	2010	Informative	SFA [2010] No. 26

Major national climate policies in China since 2000 (continued)

Sector	Policy	Dates	Type	Issuing ministries and policy number
	China's Forestry Action Plan to Deal with Climate Change This action plan stipulates the goals of three stages. By 2050, the forest area will realize a net growth of 47 million hectares compared with 2020, and the carbon-sequestering ability of forests of the country will reach a stable level. The plan stipulates twenty-two major actions of the forestry industry, including fifteen actions for mitigating climate change and seven actions for adapting to it.	2009	Plan	Serial number unknown
Residential and commercial	**The 13th Five-Year Plan for the Development of Energy-Efficient and Green Building** The plan objectives are to improve the energy-efficiency for all new urban residential and public buildings by 20 percent compared with 2015, while building energy conservation standards in some regions and for key architectural components, such as windows and doors, with the goal to meet or come very close to the current international advanced level by 2020. Green buildings shall account for more than 50 percent of all new urban buildings, with the use of energy-saving construction materials exceeding 40 percent. Energy-efficiency renovation needs to be carried out for more than 500 million m² of existing residential buildings and 100 million m² of public buildings. Among existing residential buildings in cities and towns across the country, at least 60 percent shall achieve green building standards.	2017, plan for the previous five-year plan period issued in 2012	Plan	MOHURD [2017] No. 53
	The Solar Energy for Poverty Alleviation Programme The program aims to add over 10 GW capacity and benefit more than two million households from around thirty-five thousand villages across the country by 2020.	Updated in 2016, first introduced in 2014	Fiscal	NDRC Energy [2016] No. 621

Name	Description	Date	Type	Code
Design Standards and Acceptance Codes for Residential and Public Buildings	China's first energy-efficient building design standard was implemented in 1986 for buildings in the northern heating zones, with the standard being updated in 1995 and again in 2010. China also issued the Code for Acceptance of Energy Efficient Building Construction in 2007. The acceptance code makes compliance with building energy-efficiency requirements mandatory for the final acceptance of a construction project.	Updated in 2015, first introduced in 1986	Regulatory	GB50189–2015
Green Building Evaluation Standard (the Three Star System)	The system grants three levels, with a three stars level marking the highest-rated green buildings. The updated green building standard provides evaluation protocols to differentiate between residential buildings and public-purpose buildings and to provide bonus points for ongoing improvements.	Revised in 2014, first introduced in 2006	Informative Voluntary	GB—T50378–2014
Standard for Lighting Design in Buildings	For some building types, the maximum lighting power density values defined by the new Chinese standard are slightly lower than values defined by the Building Area Method in ASHRAE 90.1–2013 of the United States.	Updated in 2013, first introduced in 2004	Regulatory	GB 50034–2013
Green Building Action Plan	The plan puts forward the goal of completing 1 billion m^2 of new green buildings during the period of the twelfth five-year plan and demanding the full implementation of green building standards from 2014 for government-invested buildings, public housing in major cities, and large public-purpose buildings with an area of more than 20,000 m^2.	2013	Plan	No. 1 [2013] of the General Office of the State Council

Major national climate policies in China since 2000 (continued)

Sector	Policy	Dates	Type	Issuing ministries and policy number
	Implementation Guidance on Accelerating the Development of Green Buildings in China The guidance specifies that two-star green buildings enjoy a subsidy of RMB 45 per m² and three-star green buildings enjoy a subsidy of RMB 80 per m² (construction area).	2012	Fiscal	MOF & MOHURD [2012] No. 167
	Roadmap for the Phasing-Out of Incandescent Bulbs The roadmap pledged to replace the one billion bulbs used annually with more energy-efficient models within five years. As of October 1, 2016, incandescent bulbs above the fifteen-watt range were required to be eliminated from all retailers, completing the five-year task.	2011	Regulatory	NDRC [2011] No. 28
	Energy Conservations in Public Institutions In 2008, the State Council issued *Regulation on Energy Conservation by Public Institutions*, a key regulation describing the responsibilities and requirements of all public institutions for improving energy efficiency. *12th FYP for Energy Conservation in Public Institutions*, published in 2011, includes the targets of reducing public institutions' energy consumption per person by 15 percent and unit energy consumption for building floor area by 12 percent by 2015 as compared to 2010.	2011 and previously in 2008	Regulatory	NGOA Energy Conservation [2011] No. 433
	Circular on Establishing System of Compulsory Government Procurement of Energy Conservation Products Up to June 2017, the NDRC and MOF have published twenty-two editions of the government procurement lists.	2007	Regulatory	No. 51 [2007] of the General Office of the State Council

Provincial or local policies

Mandatory Coal Cap Targets The Airborne Pollution Prevention and Control Action Plan unveiled by the State Council in 2013 pledged to cap coal consumption and improve the air quality of the entire country by 2017. Thereafter, six Chinese provinces have established absolute coal-consumption-reduction targets in their air pollution action plans, with a 50 percent reduction targeted in Beijing, 13 percent in Hebei, 19 percent in Tianjin, 5 percent in Shandong, 21 percent in Chongqing, and 13 percent in Shaanxi by the end of 2017, compared to 2012 levels.	Various dates between 2013 and 2017	Regulatory	No. 37 [2013] the General Office of State Council
Low-Carbon City In 2010, the Low Carbon City Pilot Scheme was formally endorsed by the NDRC. The main aims of the scheme were to develop low-CO_2-emission industries, establish a greenhouse gas emission data collection and management system, and encourage residents to adopt green consumption patterns. In 2012, a total of six provinces and thirty-six cities were involved in the scheme. The NDRC issued another forty-five low-carbon pilot cities at the beginning of 2017, thus raising the total number of low-carbon provinces and cities to eighty-seven.	Updated in 2017, first introduced in 2010	Guideline	NDRC Climate [2017] No. 66
Carbon-Emissions-Trading Pilot Programs In 2011, the NDRC approved establishing carbon-emissions-trading pilot programs in seven provinces or cities (Beijing, Chongqing, Shanghai, Tianjin, Guangdong, Hubei, and Shenzhen). Following Shenzhen launching its trading in June 2013, Shanghai, Beijing, Guangdong, and Tianjin in turn launched their first trading before the end of 2013. The remaining two of the seven pilot schemes, Hubei and Chongqing, launched trading on April 2 and June 19, 2014, respectively.	2011	Market	NDRC [2011] No. 2601

Sources: Each of these policies were directly sourced from Chinese government websites. All of the Chinese sources were documented on the Chinese-language website (and the relevant policy number is provided) and then translated for the purposes of this table.

Major national climate policies in the United States

Sector	Policy	Dates	Type	Implementing agencies
Economy wide				
	Climate Action Plan Includes three sections: cutting carbon pollution in America, preparing the United States for the impacts of climate change, and leading international efforts to address global climate change. Puts the United States on track to meet its 2020 target and lay the foundation to reach the 2025 target.	2013	Plan	Issued by the executive office of former president Obama
	Climate Action Plan—Strategy to Cut Methane Emissions Outlines new actions to reduce methane emissions. Steps along these lines could deliver greenhouse gas emissions reductions up to 90 million metric tons in 2020.	2013	Plan	Issued by the executive office of former president Obama
	GHG emission reporting programs Mandates reporting of greenhouse gases from sources that in general emit twenty-five thousand metric tons or more of carbon dioxide equivalent per year in the United States.	2009	Information	EPA
	National Action Plan for Energy Efficiency Creates a sustainable, aggressive national commitment to energy efficiency through the collaborative efforts of gas and electric utilities, utility regulators, and other partner organizations. The goal is to achieve all cost-effective energy efficiency measures by 2025.	2006	Plan	EPA

Transportation

	Year	Type	Agency
GHG Emissions and Fuel-Efficiency Standards for Medium- and Heavy-Duty Engines and Vehicles in model years 2018–2027 To lower carbon emissions by approximately one billion metric tons, cut fuel costs by about USD 170 billion, and reduce oil consumption by up to 1.8 billion barrels over the lifetime of the vehicles sold under the program.	2016	Regulatory	EPA and NHTSA
Renewable Fuel Standard Mandates the deployment and use of thirty-six billion gallons of renewable fuel annually by 2022.	2005; amended in 2007 and 2015	Regulatory	EPA
EV Everywhere Grand Challenge To produce plug-in electric vehicles (PEVs) that are as affordable and convenient for the American family as gasoline-powered vehicles by 2022. The challenge includes R&D strategy, consumer education, and PEV-ownership incentives.	2012	Fiscal Innovation	DOE
Corporate average fuel-economy standards for passenger cars and light trucks for model years (MYs) 2017 through 2025 To raise the average new vehicle fuel efficiency of cars and light trucks to roughly 54.5 mpg and lower the CO_2 emission level to a compliance-based value of 163 gpm in 2025	2012	Regulatory	EPA

Major national climate policies in the United States (continued)

Sector	Policy	Dates	Type	Implementing agencies
	GHG emissions standards and fuel-efficiency standards for medium- and heavy-duty engines and vehicles in MYs 2014 to 2018 To reduce fuel consumption by 6 to 23 percent in the MY 2017 timeframe, as compared to a MY 2010 baseline, and reduce oil consumption by 530 million barrels and GHG pollution by approximately 270 Mt over the lifetime of the vehicles.	2011	Regulatory	EPA and NHTSA
	Standards for light-duty passenger vehicles including passenger cars and light trucks for MYs 2012 to 2016 To raise the average new vehicle fuel economy to over 35.5 miles per gallon (mpg) and result in average new vehicle tailpipe CO_2 emissions of 250 grams per mile (gpm) by MY 2016	2010	Regulatory	EPA
Energy—Power Supply	**Competitive Processes, Terms, and Conditions for Leasing Public Lands for Solar and Wind Energy Development and Technical Changes and Corrections** To facilitate responsible solar and wind energy development on BLM-managed public lands and to ensure that the American taxpayer receives fair market value for such development.	2016	Regulatory Fiscal	BLM

Policy	Date	Type	Agency
Clean Power Plan—"Carbon Pollution Emission Guidelines for Existing Stationary Sources: Electric Utility Generating Units" To reduce carbon dioxide emissions from electrical power generation by 32 percent from 2005 levels by 2030.	Rules finalized in 2015; enforcement halted by the Supreme Court in 2016; currently under review	Regulatory	EPA
Federal Renewable Production Tax Credit (PTC) An inflation-adjusted per-kilowatt-hour (kWh) tax credit for electricity generated by qualified energy resources and sold by the taxpayer to an unrelated person during the taxable year. The duration of the credit is ten years after the date the facility is placed in service for all facilities placed in service after August 8, 2005. The credit differs by type of renewable energy.	Enacted in 1992; renewed and expanded many times, most recently in December 2015	Fiscal	IRS
Business Energy Investment Tax Credit (ITC) The credit was most recently amended by the Consolidated Appropriations Act of 2015, which extended the expiration date, but also introduced a step-down in the value of the credit for solar technologies and PTC-eligible wind between 2019 and 2022.	Enacted in 1978 and amended a number of times, most recently in December 2015	Fiscal	IRS
Clean Energy Investment Initiative The initiative has leveraged more than USD 4 billion in commitments from foundations, institutional investors, and philanthropies to fund solutions to help fight climate change.	2015	Innovation	DOE

Major national climate policies in the United States (continued)

Sector	Policy	Dates	Type	Implementing agencies
	Rural Energy for America Program—Renewable Energy Systems and Energy-Efficiency Improvement Loans and Grants Provides guaranteed loan financing and grant funding to agricultural producers and rural small businesses for renewable energy systems or to make energy-efficiency improvements	First enacted in 2003; reauthorized in 2013	Fiscal	USDA
	Energy-Efficiency and Conservation Loan Program Provides loans to finance energy-efficiency and conservation projects for commercial, industrial, and residential consumers. These funds also can be used for distributed generation for on- or off-grid renewable energy systems.	2013	Fiscal	USDA
	SunShot initiative To make solar power fully cost-competitive with traditional energy sources, without incentives, by 2020.	2011	Fiscal	DOE
	US Department of Energy—Loan Guarantee Program Issues loan guarantees for innovative energy projects and loans for advanced technology vehicle manufacturing projects. LPO manages a portfolio comprising more than USD 30 billion of loans, loan guarantees, and conditional commitments covering more than thirty projects. Overall, these loans and loan guarantees have resulted in more than USD 50 billion in total project investment.	Created in 2005; reauthorized in 2009 and 2011	Fiscal	DOE

Program	Date	Type	Agency
Advanced Research Projects Agency—Energy As of January 2015, the program has funded more than four hundred high-potential, high-impact energy projects through twenty-five focused programs and open-funding solicitations.	Created in 2007; first batch of funding received in 2009	Innovation	DOE
Enhancing Renewable Energy Development on the Public Lands Facilitates the Department of the Interior's efforts to approve nonhydropower renewable energy projects on the public lands with a generation capacity of at least ten thousand megawatts of electricity by 2015.	2009	Regulatory	BLM
Energy: residential, commercial, and industrial end use			
Appliance, Equipment, and Lighting Energy Efficiency Standards Establishes minimum energy conservation standards for more than sixty categories of appliances and equipment. Products covered by the standards represent about 90 percent of home energy use, 60 percent of commercial building use, and 30 percent of industrial energy use.	Established in 1975; amended several times, including modifications in 2012	Regulatory Informative	DOE
Better Buildings To improve the lives of the American people by driving leadership in innovation in building energy efficiency.	2011	Voluntary Innovation	DOE
Better Buildings Challenge As part of the Better Buildings initiative, the challenge is aimed at achieving the goal of doubling American energy productivity by 2030 while motivating corporate and public-sector leaders across the country to save energy through commitments and investments.			

Major national climate policies in the United States (continued)

Sector	Policy	Dates	Type	Implementing agencies
	ENERGY STAR	1992; amended several times	Informative Voluntary	EPA
	Identifies and promotes energy-efficient products, now inclusive of major appliances, office equipment, lighting, home electronics, new homes, and commercial and industrial buildings and plants.			
	Covers products in more than seventy different categories, with more than 4.8 billion sold since 1992. More than 1.5 million new homes and more than twenty-two thousand facilities carry EPA's ENERGY STAR certification.			
	Building Energy Codes Program	Initially established in 1992; updated regularly	Regulatory	DOE
	Sets minimum efficiency requirements for newly constructed and renovated buildings with adoption and compliance strategies.			
Industrial	**New Source Performance Standards— Permitting Rules for the Oil and Natural Gas Industry**	2012; updated in 2016	Regulatory	EPA
	The 2016 updates to the NSPS include standards for methane and volatile organic compound (VOC) emissions from new, reconstructed, and modified oil and gas sources. The rule is expected to prevent emissions equal to 7.7–9.0 million Mt CO_2e by 2025. The benefits of these actions are estimated at USD 120–150 million in 2025.			

Significant New Alternatives Policy (SNAP) To identify and evaluate substitutes in end uses that have historically used ozone-depleting substances; look at overall risk to human health and the environment of both existing and new substitutes; publish lists of acceptable and unacceptable substitutes by end use; promote the use of acceptable substitutes; and provide the public with information about the potential environmental and human health impacts of substitutes. The Obama Administration finalized a rule in July 2015 to prohibit some of the most harmful HFCs in various end uses under SNAP. In 2016, EPA published final rules listing new substitutes and prohibiting certain high-GWP HFCs as alternatives under SNAP.	1994; new rules published on an ongoing basis	Regulatory Information	EPA
Coalbed Methane Outreach Program (CMOP) The program has worked cooperatively with the coal-mining industry in the United States—and other major coal-producing countries—to reduce methane emissions. By 2014, the program is estimated to have cumulatively reduced approximately 180 MMTCO$_2$e.	1994	Information	EPA

Major national climate policies in the United States (continued)

Sector	Policy	Dates	Type	Implementing agencies
Land use/adaptation				
	US Department of Agriculture's (USDA's) Building Blocks for Climate Smart Agriculture and Forestry	2015	Plan	USDA
	To help farmers, ranchers, forestland owners, and rural communities respond to climate change. Consists of the following ten "building blocks": soil health, nitrogen stewardship, livestock partnerships, conservation of sensitive lands, grazing and pasture lands, private forest growth and retention, stewardship of federal forests, promotion of wood products, urban forests, and energy generation and efficiency. Through this initiative, USDA intends to reduce GHG emissions and increase carbon stored in forests and soils by over 120 million metric tons of carbon dioxide equivalent per year by 2025.			
	Priority Agenda for Enhancing the Climate Resilience of America's Natural Resources	2014	Plan	Issued by the executive office of former president Obama
	Identifies four priority strategies to make the nation's natural resources more resilient to climate change:			
	• Foster climate-resilient lands and waters			
	• Manage and enhance US carbon sinks			
	• Enhance community preparedness and resilience by utilizing and sustaining natural resources			
	• Modernize federal programs, investments, and delivery of services to build resilience and enhance sequestration of biological carbon			

AgSTAR Supports farmers and industry in the development and adoption of anaerobic-digester systems—specialized manure-management systems that capture biogas. Cumulatively, anaerobic digesters on livestock farms are estimated to have reduced emissions by 5.6 Mt CO_2e since 2000.	1994	Voluntary	EPA
Federal Government			
Planning for Federal Sustainability in the Next Decade Establishes overall agency GHG emissions goals and associated planning requirements, which include GHG emission reductions from federal fleets. Aims to reduce GHG emissions in the federal government by 40 percent by 2025 and to increase the share of electricity the federal government consumes from renewable and alternative sources to 30 percent.	2015	Plan	DOE
Energy-Efficiency and Renewable Energy Initiatives—Department of Defense To reduce its fossil fuel use by improving energy efficiency (i.e., reducing wasted energy) and shifting to renewable energy, with a goal of producing 25 percent of its energy from renewable sources by 2025.	2010	Plan	DOD

Sources: Each of these policies were directly sourced from US government websites.

Notes

Foreword by John P. Holdren

1. Richard J. Millar, Jan S. Fuglestvedt, Pierre Friedlingstein, Joeri Rogelj, Michael J. Grubb, H. Damon Matthews, Ragnhild B. Skeie, Piers M. Forster, David J. Frame, and Myles R. Allen, "Emissions Budgets and Pathways Consistent with Limiting Warming to 1.5 C," *Nature Geoscience* 10 (2017): 741–747, https://doi.org/10.1038/NGEO3031.

2. For 2016 (the latest data available), Chinese CO_2 emissions from industrial sources accounted for 29 percent of the world's total, and US emissions accounted for 14 percent. See http://edgar.jrc.ec.europa.eu/overview.php?v=CO2andGHG1970-2016&sort=des8.

1 Introduction

1. Two members of the US foreign policy establishment were exceptions to this rule: former vice president Al Gore and the secretary of state at the time, John Kerry. Both former senators, the two men had championed climate change as a foreign policy matter later in their careers.

2. Defined by Merriam-Webster as "one that is gigantic in size and power."

3. As discussed in chapter 4.

4. Chinese economic development strategy is discussed in greater detail in chapter 2.

5. This memorandum was submitted by one of the authors of this book, Kelly Sims Gallagher. While senior officials in the Obama administration embraced the overall concept of a bilateral agreement, some aspects of her recommendations were not adopted at the time. She advocated a bilateral agreement on domestic policies to reduce emissions that was not necessarily situated within the UNFCCC.

6. The technical exchange was coordinated by Dr. Zou Ji (邹骥) on the Chinese side and Gallagher on the US side and included experts from government agencies in the EPA, DOE, State Department, and White House on the US side and experts from the NCSC, Chinese Academy of Social Sciences, Chinese Academy of Engineering, Tsinghua University on the Chinese side.

7. China's Ministry of Foreign Affairs translates the term as "a new model of major country relationship," but most US officials use the term *major power relations*. In Chinese, the literal translation is indeed *large country* (大国), but the interpretation outside China has generally been *great power*.

8. For details on all the subnational commitments, see https://ccwgsmartcities.lbl .gov/declaration.

9. We focus exclusively on mainland China in this book and do not include any discussion of Hong Kong or Macao, both considered special administrative regions. We also do not analyze climate policy in Taiwan.

10. Twenty US states and more than four hundred US cities signed America's Pledge, according to https://www.americaspledgeonclimate.com/.

2   National Circumstances

1. All reserves data from BP 2017.

2. See http://environment.law.harvard.edu/postelection/.

3   Comparing Policymaking in Structures, Actors, Processes, and Approaches

1. These numbers of retweets and likes were as of July 20, 2017.

2. The reason that NOAA is part of the DOC is because when NOAA was created, President Richard Nixon was angry with his secretary of the interior, Wally Hickel, for being critical of his Vietnam War strategy, so Nixon decided to give NOAA to the DOC to punish Mr. Hickel (Mervis 2012).

3. Throughout this book, we focus almost exclusively on policymaking in mainland China, not including Taiwan or Hong Kong.

4. The People's Congress of an autonomous ethnic area has the power to enact autonomous decree and special decree in light of its ethnic political, economic, and cultural characteristics (Legislation Law: Article 66).

5. See http://www.cppcc.gov.cn/.

6. In 2013, the Ministry of Health (卫生部) was reorganized as the National Health and Family Planning Commission (卫生和计划委员会 or 卫计委).

7. In a Chinese context, an *interest group* can be any group that acts on behalf of a subordinate's interests. Therefore, enterprises and some bureaucracies in the government can be regarded as interest groups if they consider only their own interests. In fact, there are few interest groups in China defined similarly to how they are conceived in the US context.

8. The proposal to set up a national gas pipeline company was submitted many years ago, but it only recently made significant progress because PetroChina objected to it. The anticorruption movement in PetroChina finally caused PetroChina's views to evolve. See http://finance.sina.com.cn/chanjing/gsnews/2017-09-28/doc-ifymk wwk6789568.shtml and http://news.bjx.com.cn/html/20140526/513340.shtml for more information.

9. According to the Party Constitution (党章), the party follows the principle of *democratic centralism*, which is a combination of centralism on the basis of democracy and democracy under centralized guidance. But it is not clear which issues utilize democratic approaches and which issues are subject to centralized guidance. Usually in the meeting of a party committee or party group (党委会议 or 党组会议) at different levels and in different places, there is a vote for an important decision, such as whether seven or nine members of the Standing Committee of the Political Bureau are required for a vote. But in reality, the party secretary has absolute power in many cases and will sometimes make the final decision, after which the vote becomes simply procedural.

4 National Target Formation

1. The NDC registry can be found at http://unfccc.int/focus/ndc_registry/items/9433.php (UNFCCC 2016a).

2. For a review, see Mattoo and Subramanian 2012.

3. This method was developed by Vance Wagner and Kelly Sims Gallagher for the Obama administration's use in comparing the US, Chinese, and European NDCs prior to the Paris Agreement. Wagner served in the State Department's Office of the Special Envoy on Climate Change as China counsellor.

4. In the Chinese language, *jihua* (计划) corresponds to the planned economy. The meaning of the word is that the government makes the plan and decides how the economy will work, such as how much to invest and produce. *Guihua* (规划) corresponds to the market economy. The meaning of guihua is that the market decides how the economy will work, but the government will still provide guidelines and set some targets for the economy. These targets include two different types. One type is the *compulsory target* (强制性指标), which means the government requires achievement of this target, such as energy-intensity reductions. The other type is the *anticipated target* (预期性指标), which is only a forecast, such as the economic growth rate.

5 Target Implementation

1. For more information, see http://history.house.gov/Institution/Origins-Develop
ment/Power-of-the-Purse/.

2. After one of the Supreme Court justices, Antonin Scalia, died at the end of the
Obama term, the Senate Republican majority leader, Mitch McConnell, refused to
hold a vote on President Obama's new Supreme Court nominee, in hopes that a
Republican would be elected president and then nominate a more conservative
judge. McConnell's refusal to allow for the vote was unprecedented historically and
is a good example of the extreme partisanship in the US Congress as of 2017. See
https://www.politico.com/story/2016/11/supreme-court-mitch-mcconnell-231150.

3. See http://www.mlr.gov.cn/mlrenglish/about/mission/.

6 Why Climate Policy Outcomes Differ

1. A video with a link to the advertisement is available here: https://www.huffing
tonpost.com/2010/10/11/joe-manchin-ad-dead-aim_n_758457.html.

2. See https://www.affordablesolarusa.org/.

3. See http://www.americansolarmanufacturing.org/news-releases/05-17-12
-commerce-department-ruling.htm.

7 Conclusion

1. The State Planning Commission was founded in 1952, and its name was changed
to the State Development Planning Commission (SDPC) in 1998. In 2003, the name
was changed again to the National Development and Reform Commission (NDRC).

2. The elected Senators from each party determine committee membership before
the start of each new Congress. Based on the last election, the percentage of seats
allocated to each party on each committee is based on the percentage of a party's
representation in the Senate. Traditionally, the senator with the greatest seniority
on a particular committee serves as its chairman; however, when the Republican
party gained the majority in 1995, it altered its rules to allow Republicans to vote by
secret ballot for their committee's chairman, irrespective of that member's seniority
(US Senate 2017).

References

Allison, Graham. 2017. *Destined for War: Can America and China Escape the Thucydides Trap?* Boston: Houghton Mifflin Harcourt.

Allison, Graham T., and Morton H. Halperin. 1972. "Bureaucratic Politics: A Paradigm and Some Policy Implications." *World Politics* 24:40–79.

Alvarez, R., S. Pacala, J. Winebrake, W. Chameides, and S. Hamburg. 2012. "Greater Focus Needed on Methane Leakage from Natural Gas Infrastructure." *Proceedings of the National Academy of Sciences of the United States of America* 109 (12): 6435–6440.

Ansolabehere, Stephen, and David M. Konisky. 2012. "The American Public's Energy Choice." *Daedalus* 141 (2): 61–71.

APPC (Alliance of Peaking Pioneer Cities). 2017. Alliance of Peaking Pioneer Cities of China website. http://appc.ccchina.gov.cn.

Archives (National Archives and Records Administration). 2017. "What Is the Electoral College?" https://www.archives.gov/federal-register/electoral-college/about.html.

Barnett, A. Doak. 1985. *The Making of Foreign Policy in China*. Boulder: Westview Press.

Barradale, Merrill J. 2010. "Impact of Public Policy Uncertainty on Renewable Energy Investment: Wind Power and the Production Tax Credit." *Energy Policy* 38 (2): 7698–7709.

Baumgartner, Frank R., Bryan D. Jones, and Peter B. Mortensen. 2014. "Punctuated Equilibrium Theory: Explaining Stability and Change in Public Policymaking." In *Theories of the Policy Process*, 3rd ed., ed. Christopher M. Weible and Paul A. Sabatier, 59–104. Boulder, CO: Westview Press.

BBC News. 2015. "China Pollution: First-Ever Red Alert in Effect in Beijing." *BBC.com*, December 8.

Bottemiller, Hellena. 2010. "China Launches Food Safety Commission." *Food Safety News*, February 11. http://www.foodsafetynews.com/2010/02/china-launches-food -safety-commission/#.WZ5vtGP9wWd.

Boykoff, Maxwell. 2007. "Flogging a Dead Norm? Newspaper Coverage of Anthropogenic Climate Change in the United States and United Kingdom from 2003 to 2006." *Area* 39 (4): 470–481.

BP (British Petroleum). 2015. "Statistical Review of World Energy." *British Petroleum*. https://www.bp.com/en/global/corporate/energy-economics/statistical-review-of -world-energy.html.

BP (British Petroleum). 2017. "Statistical Review of World Energy." *British Petroleum*. https://www.bp.com/en/global/corporate/energy-economics/statistical-review-of -world-energy.html.

Broder, John. 2013. "Obama Readying Emissions Limits on Power Plants." *New York Times*, June 19.

Bruce-Lockhart, Anna. 2017. "Top Quotes by China President Xi Jinping at Davos 2017." *World Economic Forum*, January 17. https://www.weforum.org/agenda/2017/ 01/chinas-xi-jinping-at-davos-2017-top-quotes/.

BLS (US Bureau of Labor Statistics). 2017. "Data Retrieval: Employment, Hours, and Earnings (CES)." *United States Department of Labor, Bureau of Labor Statistics*, February 2. https://www.bls.gov/webapps/legacy/cesbtab1.htm.

BLS (US Bureau of Labor Statistics). 2018. "Data Retrieval: Employment, Hours, and Earnings (CES)." *United States Department of Labor, Bureau of Labor Statistics*, February 2. https://www.bls.gov/webapps/legacy/cesbtab1.htm.

Byrd-Hagel Resolution. 1997. US Senate Res. 98. 105th Cong. Sess. 143th Congressional Record (July 25).

C2ES. 2002. "Analysis of President Bush's Climate Change Plan." Center for Climate and Energy Solutions [online database]. https://www.c2es.org/document/analysis-of -president-bushs-climate-change-plan/.

CECC (Congressional-Executive Commission on China). n.d. "China's State Organizational Structure." http://www.cecc.gov/chinas-state-organizational-structure. Accessed December 30, 2017.

Center for Responsive Politics. 2017. "Interest Groups." *OpenSecrets.org*. https://www .opensecrets.org/industries/. Accessed August 20, 2017.

Central People's Government of the People's Republic of China. 2016. *The 13th Five-Year Plan for Economic and Social Development of the People's Republic of China (2016–2020)*. Beijing: National People's Congress and Chinese People's Political Consultative Congress.

Cheng, Xiaonong. 2016. "Capitalism with Chinese Characteristics: From Socialism to Capitalism." *Epoch Times*, July 10. https://www.theepochtimes.com/n3/2111687 -capitalism-with-chinese-characteristics-from-socialism-to-capitalism/.

Chi, Ma. 2016. "Government Posts Remain Appealing to Job Seekers: Experts." *China Daily Online*, June 21, 2016. http://www.chinadaily.com.cn/china/2016-06/ 21/content_25787206.htm.

China Daily. 2017. "Xi Leads Ecological Civilization." *China Daily*, March 22, 2017.

China.org.cn. n.d. "The Basic Tasks Set Out in the 10th Five-Year Plan (2001–2005)." http://www.china.org.cn/english/MATERIAL/157629.htm. Accessed December 30, 2017.

ChinaNews. 2013. "The New Leadership Lineup of the Commission of Political and Legal Affairs under the CPC Central Committee." *ChinaNews.com*, April 8. http:// www.chinanews.com/gn/2013/04-08/4710970.shtml.

Clifford, Mark. 2015. "Can China's Top-Down Approach Fix Its Environmental Crisis?" *The Guardian*, June 4. https://www.theguardian.com/sustainable-business/ 2015/jun/04/can-chinas-top-down-approach-fix-its-environmental-crisis.

CMA (China Meteorological Administration). 2011. "The Third National Climate Change Expert Committee Was Established and Held Its First Working Meeting." http://www.cma.gov.cn/2011xwzx/2011xqxxw/2011xqxyw/201609/t20160930 _325288.html.

Cohen, Michael D., James G. March, and Johan P. Olsen. 1972. "A Garbage Can Model of Organizational Choice." *Administrative Science Quarterly* 17 (1): 1–25.

Davenport, Coral. 2015. "A Climate Deal, 6 Fateful Years in the Making." *New York Times*, December 13.

Davenport, Coral. 2017. "Counseled by Industry, Not Staff, E.P.A. Chief Is Off to a Blazing Start." *New York Times*, July 1.

Davis, Steven, and Ken Caldeira. 2010. "Consumption-Based Accounting of CO_2 Emissions." *Proceedings of the National Academy of Sciences of the United States of America* 107 (12): 5687–5692.

Dembicki, Geoff. 2017. "The Convenient Disappearance of Climate Change Denial in China." *Foreign Policy* (May/June): 31.

DeSombre, Elizabeth R. 2000. *Domestic Sources of International Environmental Policy: Industry, Environmentalists, and US Power*. Cambridge, MA: MIT Press.

DOE (US Department of Energy). n.d-a. "Business Energy Investment Tax Credit." https://energy.gov/savings/business-energy-investment-tax-credit-itc. Accessed December 30, 2016.

DOE (US Department of Energy). n.d.-b. "DOE Loan Programs Office Portfolio." https://energy.gov/lpo/portfolio. Accessed December 30, 2016.

DOE (US Department of Energy). n.d.-c. "FY18 Budget Justification to Congress." https://energy.gov/cfo/downloads/fy-2018-budget-justification. Accessed December 30, 2016.

DOE (US Department of Energy). n.d.-d. "Renewable Electricity Production Tax Credit." https://energy.gov/savings/renewable-electricity-production-tax-credit-ptc. Accessed December 30, 2016.

DOE (US Department of Energy). n.d.-e. "Residential Renewable Energy Tax Credit." https://energy.gov/savings/residential-renewable-energy-tax-credit. Accessed December 30, 2017.

DOE DSIRE (Database of State Incentives for Renewables and Efficiency). 2016. "Renewable Generation Requirement." DSIRE. Last updated April 29, 2016. http://programs.dsireusa.org/system/program/detail/182.

Dong, Jun, Yu Ma, and Hongxing Sun. 2016. "From Pilot to the National Emissions Trading Scheme in China: International Practice and Domestic Experiences." *Sustainability* 8 (6): 1–17.

DRC (Development Research Center of the State Council). 2013. "Message from the President." DRC, August 29, 2013. http://en.drc.gov.cn/2013-08/29/content _16955102.htm.

E&E News Reporter. 2017. "Your Guide to the Clean Power Plan in the Courts." *E&E News*, sec. Power Plan Hub.

Eaton, Sarah, and Genia Kostka. 2014. "Authoritarian Environmentalism Undermined? Local Leaders' Time Horizons and Environmental Policy Implementation in China." *China Quarterly* 218:359–380.

Economy, Elizabeth. 2014. "Environmental Governance in China: State Control to Crisis Management." *Daedalus* 143 (2): 184–197.

EIA (US Energy Information Administration). 2016a. *Annual Energy Outlook 2016*. Washington, DC: US Department of Energy.

EIA (US Energy Information Administration). 2016b. "Electric Power Monthly: Net Generation from All Renewable Sources." https://www.eia.gov/electricity/monthly/archive/august2016.pdf.

EIA (US Energy Information Administration). 2016c. *International Energy Outlook 2016*. Washington, DC: US Department of Energy.

EIA (US Energy Information Administration). 2018a. *Consumption and Efficiency Data Browser*. Washington, DC: US Department of Energy. https://www.eia.gov/consumption/data.php. Accessed July 3, 2018.

EIA (US Energy Information Administration). 2018b. "State Carbon Dioxide Emissions Database." https://www.eia.gov/environment/emissions/state/. Accessed July 3, 2018.

Eilperin, Juliet, and Brady Dennis. 2017. "EPA to Pull Back on Fuel-Efficiency Standards for Cars, Trucks in Future Model Years." *Washington Post*, March 3.

EPA (US Environmental Protection Administration). 2016. "Regulations for Greenhouse Gas Emissions from Passenger Cars and Trucks." https://www.epa.gov/regulations-emissions-vehicles-and-engines/regulations-greenhouse-gas-emissions-passenger-cars-and.

EPA (US Environmental Protection Administration). 2017a. "The Basics of the Regulatory Process." https://www.epa.gov/laws-regulations/basics-regulatory-process.

EPA (US Environmental Protection Administration). 2017b. "Enforcement Basic Information." https://www.epa.gov/enforcement/enforcement-basic-information.

EPA (US Environmental Protection Administration). 2017c. "Frequently Asked Questions about Trailer Standards for Fuel Efficiency and Greenhouse Gas Emissions." https://nepis.epa.gov/Exe/ZyPURL.cgi?Dockey=P100QWHL.TXT.

EPA (US Environmental Protection Administration). 2017d. "How the Energy Independence and Security Act of 2007 Affects Light Bulbs." https://www.epa.gov/cfl/how-energy-independence-and-security-act-2007-affects-light-bulbs.

EPA (US Environmental Protection Administration). 2017e. "Mercury and Air Toxics Standards." https://www.epa.gov/mats/cleaner-power-plants.

EPA (US Environmental Protection Administration). 2018. *Inventory of US Greenhouse Gas Emissions and Sinks: (1990–2016)*. Washington, DC: EPA.

EPA Archive. 2015a. "Factsheet: The Clean Power Plan: By the Numbers." https://archive.epa.gov/epa/sites/production/files/2015-08/documents/fs-cpp-by-the-numbers.pdf.

EPA Archive. 2015b. "Factsheet: The Clean Power Plan: Key Changes and Improvements." https://archive.epa.gov/epa/sites/production/files/2015-08/documents/fs-cpp-key-changes.pdf.

EPA Archive. 2017. "Clean Power Plan for Existing Power Plants: Regulatory Actions." https://archive.epa.gov/epa/cleanpowerplan/clean-power-plan-existing-power-plants-regulatory-actions.html.

Erikson, R. S. 2001. "The 2000 Presidential Election in Historical Perspective." *Political Science Quarterly* 116 (1): 29–52.

Everett, Burgess. 2016. "McConnell's Supreme Court Gamble Pays Off in Spades." *POLITICO Magazine*, November 10.

Fairbank, John King. 1979. *The United States and China.* 4th ed. Cambridge, MA: Harvard University Press.

Federal Reserve Bank of Saint Louis. 2018. "Federal Net Outlays as Percent of Gross Domestic Product." https://fred.stlouisfed.org/series/FYONGDA188S.

Feng, K., S. Davis, L. Sun, and K. Hubacek. 2015. "Drivers of the US CO2 Emissions 1997–2013." *Nature Communications* 6 (7714). doi:10.1038/ncomms8714.

Fewsmith, Joseph. 2004. "Promoting the Scientific Development Concept." *China Leadership Monitor,* July 30, 2004.

Fewsmith, Joseph. 2013. *The Logic and Limits of Political Reform in China.* New York: Cambridge University Press.

Figueres, Christiana. 2016. "Remarks at the Fletcher School." Lecture at Tufts University, Medford, MA, April 7.

Fransen, Taryn, Juan-Carlos Altamirano, Heather McGray, and Kathleen Mogelgaard. 2015 "Mexico Becomes First Developing Country to Release New Climate Plan (INDC)." *World Resources Institute,* March 31. http://www.wri.org/blog/2015/03/mexico-becomes-first-developing-country-release-new-climate-plan-indc.

Freeman, Jody. 2016. "Implications of Trump's Victory and the Republican Congress for Environmental, Climate and Energy Regulation: Not as Bad as It Seems?" *Harvard Environmental Policy Initiative,* November 10.

Friedman, Lisa. 2017. "Court Blocks E.P.A. Effort to Suspend Obama-Era Methane Rule." *New York Times,* July 3.

Gallagher, Kelly S. 2006. *China Shifts Gears: Automakers, Oil, Pollution, and Development.* Cambridge, MA: MIT Press.

Gallagher, Kelly S. 2014. *The Globalization of Clean Energy Technology: Lessons from China.* Cambridge, MA: MIT Press.

Gallagher, Kelly S., and Laura D. Anadon. 2017. "DOE Budget Authority for Energy Research, Development, and Demonstration Database." The Fletcher School, Tufts University; and Cambridge University. https://sites.tufts.edu/cierp/publications/#2017.

Gallagher, Kelly S., Fang Zhang, and Robbie Orvis. 2018. "A Policy Gap Analysis for China's Climate Targets in the Paris Agreement." Unpublished paper.

Garcia, Eduardo. 2017. "California Assembly Bill 398." California State Legislature. http://www.climatechange.ca.gov/state/legislation.html.

Ge, Mengpin, Johannes Friedrich, and Thomas Damassa. 2014. "6 Graphs Explain the World's Top 10 Emitters." World Resources Institute, November 25. http://www.wri.org/blog/2014/11/6-graphs-explain-world's-top-10-emitters.

General Office of the State Council of the People's Republic of China. 2016. "Notice on Printing and Distributing the State Council's 2016 Legislative Work Plan." State Council, Beijing, April 13, 2016. http://www.gov.cn/zhengce/content/2016-04/13/content_5063670.htm.

Gilens, Martin, and Benjamin I. Page. 2014. "Testing Theories of American Politics: Elites, Interest Groups, and Average Citizens." *Perspectives on Politics* 12 (3) (September): 564–581.

Gilley, Bruce. 2012. "Authoritarian Environmentalism and China's Response to Climate Change." *Environmental Politics* 21 (2): 287–307.

Gordon, Robert. 2016. *The Rise and Fall of American Growth*. Princeton, NJ: Princeton University Press.

Government of the People's Republic of China. 2016. *The People's Republic of China First Biennial Update Report on Climate Change*. Bonn: UNFCCC.

Greene, D. L. 1998. "Why CAFE Worked." *Energy Policy* 26 (8): 595–613.

Greenhouse, L. 2007. "Justices Say E.P.A. Has Power to Act on Harmful Gases." *New York Times*, April 3.

GSA (US General Services Administration). 2000. "Public Access to Advisory Committee Records." https://www.gsa.gov/portal/content/100785.

GSA (US General Services Administration). 2017. "The Federal Advisory Committee Act (FACA) Brochure." https://www.gsa.gov/portal/content/101010.

Gu, Zhenqiu. 2015. "Interview: UN Chief Lauds China's 'Constructive, Active Role' in Addressing Climate Change." *Xinhua News Agency*, November 29.

Guangdong. 2017. "Notice on the Implementation of Measures to Control Greenhouse Gas Emissions." People's Government of Guangdong Province. May 5. http://zwgk.gd.gov.cn/006939748/201705/t20170524_706623.html.

Gupta, Joyeeta. 2010. "A History of International Climate Change Policy." *WIRES: Climate Change* 1 (5): 636–653.

Hamilton, Alexander, James Madison, and John Jay. 1961. *The Federalist Papers*. New York: New American Library of World Literature, Inc.

Hart, Craig, Jiayan Zhu, Jiahui Ying, and the Renmin University of China. 2015. *Mapping China's Policy Formation Process*. New York: Development Technologies International.

He, Zengke. 2014. "Building a Modern National Integrity System." In *China's Political Development: Chinese and American Perspectives*, ed. Kenneth G. Lieberthal, Cheng Li, and Keping Yu, 366–386. Washington, DC: Brookings Institution Press.

HLS (Harvard Law School). 2018. "Environmental Regulation Regulatory Tracker." http://environment.law.harvard.edu/policy-initiative/regulatory-rollback-tracker/.

Holden, Emily. 2017. "Pruitt Will Launch Program to 'Critique' Climate Science." *E&E News*, June 30.

Hu, An'gang. 2016. "The 13th FYP: Leading a Green Revolution." *Environmental Economy* 172–173:23–27.

Hu, An'gang, and Qingyou Guan. 2008. *Tackling Climate Change in China*. Beijing: Tsinghua University Press.

Hu, Shuli. 2008. "Tainted Milk: Regulatory Do's and Don'ts." *Caijing Magazine*, October 6.

Huang, Yasheng. 2008. *Capitalism with Chinese Characteristics*. Cambridge, UK: Cambridge University Press.

Huang, Zhenwei. 2014. "Reach a Consensus: Coordination in Decision-Making Process of China." *Journal of Hunan City University* 35 (5): 87–92.

IEA (International Energy Agency). 1999. "Coal in the Energy Supply of China." Coal Industry Advisory Board Asia Committee, OECD/IEA.

IEA (International Energy Agency). 2015. *Fossil Fuel Subsidies Database*. Paris: International Energy Agency.

IEA (International Energy Agency). 2017a. "Policies and Measures of China." OCED/IEA. https://www.iea.org/policiesandmeasures/pams/china/.

IEA (International Energy Agency). 2017b. "Statistics." OCED/IEA. http://www.iea.org/statistics/.

Inhofe, James. 2012. *The Greatest Hoax: How the Global Warming Conspiracy Threatens Your Future*. Washington, DC: WMD Books.

IPCC (Intergovernmental Panel on Climate Change). 2011. *The Fourth Assessment Report of the Intergovernmental Panel on Climate Change*. Cambridge: Cambridge University Press.

IPCC (Intergovernmental Panel on Climate Change). 2014. *The Fifth Assessment Report of the Intergovernmental Panel on Climate Change*. Cambridge: Cambridge University Press.

IRENA. 2014. *REMap 2030 China*. Abu Dhabi: International Renewable Energy Agency.

Jacobs, Lawrence R. 2013. "Lord Bryce's Curse: The Costs of Presidential Heroism and the Hope of Deliberative Incrementalism." *Presidential Studies Quarterly* 43 (4): 732–752.

Jacobson, Louis. 2014. "Mitch McConnell Says U.S.-China Climate Deal Means China Won't Have to Do Anything for 16 Years." *Politifact*, November 19. http://www.politifact.com/truth-o-meter/statements/2014/nov/19/mitch-mcconnell/mitch-mcconnell-says-us-china-climate-deal-means-c/.

Jamelske, Eric, James Boulter, Won Jang, James Barrett, Laurie Miller, and Li Han Wen. 2015. "Examining Differences in Public Opinion on Climate Change between College Students in China and the USA." *Journal of Environmental Studies and Sciences* 5 (2): 87–98.

Jamelske, Eric, James Boulter, Won Jang, James Barrett, Laurie Miller, and Li Han Wen. 2017. "Support for an International Climate Change Treaty among American and Chinese Adults." *International Journal of Climate Change: Impacts & Responses* 9 (1): 53–70.

Jenkins-Smith, Hank C., Daniel Nohrstedt, Chrisopher M. Weible, and Paul A. Sabatier. 2014. "The Advocacy Coalition Framework: Foundations, Evolution, and Ongoing Research." In *Theories of the Policy Process*, 3d ed., ed. Paul A. Sabatier and Chrisopher M. Weible, 183–224. Boulder, CO: Westview Press.

Ji, S., C. R. Cherry, M. J. Bechle, Y. Wu, and J. Marshall. 2012. "Electric Vehicles in China: Emissions and Health Impacts." *Environmental Science and Technology* 46 (4): 2018–2024.

Jin, Yuefu, Zhao Wang, Huiming Gong, Tianlei Zheng, Xiang Bao, Jiarui Fan, Michael Wang, and Miao Guo. 2015. "Review and Evaluation of China's Standards and Regulations on the Fuel Consumption of Motor Vehicles." *Mitigation and Adaptation Strategies for Global Change* 20, no. 5 (June): 735–753.

Joint Research Centre of European Commission. 2017. "CO_2 Time Series 1990–2014 per Region/Country." European Commission Joint Research Center, June 28. http://edgar.jrc.ec.europa.eu/overview.php?v=CO2ts1990-2014.

Kaiman, Jonathan. 2013. "Chinese Struggle through 'Airpocalypse' Smog." *Guardian*, February 16, 2013.

Kaiman, Jonathan. 2014. "China-US Gulf Widens as 'Marginalised' Obama Heads for Beijing Summit." *Guardian*, November 9.

Kingdon, John W. 2010. *Agendas, Alternatives, and Public Policies*. 2nd ed. Upper Saddle River, NJ: Pearson.

Kirkland, Joel. 2011. "China's Ambitious, High-Growth 5-Year Plan Stirs a Climate Debate." *New York Times*, April 12.

Kirkpatrick, David D. 2010. "Lobbyists Get Potent Weapon in Campaign Ruling." *New York Times*, January 21.

Kolbert, Elizabeth. 2015. "The Obama Administration's Self-Sabotaging Coal Leases." *New Yorker*, June 4.

Kong, Bo. 2009. *China's International Petroleum Policy*. Santa Barbara, CA: Praeger Security International.

Krauthammer, Charles. 1990. "The Unipolar Moment." *Foreign Affairs* 70 (1): 22–33.

Kroeber, Arthur. 2016. *China's Economy: What Everyone Needs to Know*. Oxford: Oxford University Press.

Kulp, Patrick. 2017. "Facebook Tweaks News Feed to Fight 'Fake News.'" *Mashable*, January 31.

Lampton, David. 1987. *Policy Implementation in Post-Mao China*. Berkeley: University of California Press.

Lampton, David. 2014a. *Following the Leader*. Berkeley: University of California Press.

Lampton, David. 2014b. "How China Is Ruled: Why It's Getting Harder for Beijing to Govern." *Foreign Affairs* 93:74.

Lawrence, Susan V. 2013. "China's Political Institutions and Leaders in Charts." Congressional Research Service, November 12, report 7-5700, R43303.

Lee, Tien Ming, Ezra M. Markowitz, Peter D. Howe, Chia-Ying Ko, and Anthony A. Leiserowitz. 2015. "Predictors of Public Climate Change Awareness and Risk Perception around the World." *Nature Climate Change* 5:1014–1020.

Legal Daily. 2006. "Energy Conservation Law Enforcement Inspection Will Start." *LegalDaily.com*, May 12. http://www.legaldaily.com.cn/misc/2006-05/12/content_313799.htm.

Levine, Mark D., Nan Zhou, and Lynn Price. 2009. "The Greening of the Middle Kingdom: The Story of Energy Efficiency in China." *The Bridge* 39 (2) (Summer): 44–54. https://www.nae.edu/19582/Bridge/EnergyEfficiency14874/14951.aspx (accessed January 12, 2011).

Lewis, Joanna. 2008. "China's Strategic Priorities in International Climate Negotiations." *Washington Quarterly* 31 (1): 155–174.

Lewis, Joanna. 2013. *Green Innovation in China: China's Wind Power Industry and the Global Transition to a Low-Carbon Economy*. New York: Columbia University Press.

Lewis, Joanna, and Ryan Wiser. 2007. "Fostering a Renewable Energy Technology Industry: An International Comparison of Wind Industry Policy Support Mechanisms." *Energy Policy* 35:1844–1857.

Li, Huimin, Ma Li, and Ye Qi. 2011. "Comparison of Climate Change Policy Processes between China and the USA." [In Chinese.] *China Population, Resources, and Environment* 21 (7): 51–56.

Li, Junfeng, Qimin Chai, Cuimei Ma, Jijie Wang, Zeyu Zhou, and Tian Wang. 2016. "China's Climate Policy and Market Outlook in the Post-Paris Era." *Energy of China* 38 (1): 5–21.

Lieberthal, Kenneth G. 1995. *Governing China: From Revolution through Reform.* 2nd ed. New York: W. W. Norton & Company.

Lieberthal, Kenneth G., Cheng Li, and Keping Yu, eds. 2014. *China's Political Development: Chinese and American Perspectives.* Washington, DC: Brookings Institution Press.

Lieberthal, Kenneth G., and Michel Oksenberg. 1988. *Policy Making in China: Leaders, Structures, and Processes.* Princeton, NJ: Princeton University Press.

Lieberthal, Kenneth G., and David Sandalow. 2009. *Overcoming Obstacles to U.S.-China Cooperation on Climate Change.* Washington, DC: Brookings Institution.

Lin, Yang, Fengyu Li, and Xian Zhang. 2016. "Chinese Companies' Awareness and Perceptions of the Emissions Trading Scheme (ETS): Evidence from a National Survey in China." *Energy Policy* 98:254–265.

Liu, Hengwei, and Kelly S. Gallagher. 2010. "Catalyzing Strategic Transformation to a Low-Carbon Economy: A CCS Roadmap for China." *Energy Policy* 38 (1): 59–74.

Liu, John Chung-En. 2015. "Low Carbon Plot: Climate Change Skepticism with Chinese Characteristics." *Environmental Sociology* 1 (4): 280–292.

Liu, John Chung-En, and Anthony A. Leiserowitz. 2009. "From Red to Green?" *Environment* 51 (4): 32–45.

Liu, Shangxi. 2010. "The Basic Idea of Further Reform of Fiscal System." *China Reform* 5: 31–37.

Lizza, Ryan. 2010. "As the World Burns: How the Senate and the White House Missed Their Best Chance to Deal with Climate Change." *New Yorker,* October 11.

Loan Programs Office of the DOE. n.d."Portfolio: Investing in American Energy." US Department of Energy. https://energy.gov/lpo/portfolio. Accessed July 3, 2017.

Loan Programs Office of the DOE. n.d."Section 1705 Loan Program." US Department of Energy. https://energy.gov/lpo/services/section-1705-loan-program. Accessed July 3, 2018.

Lou, Jiwei. 2013. *Rethinking of Intergovernmental Fiscal Relations in China.* 1st ed. Beijing: Chinese Financial & Economic Publishing House.

Lou, Jiwei. 2014. "Promoting the Standardization and Legalization of Outlay Responsibilities at All Government Levels." In *Guidance Book for the Decision of the CPC Central Committee on Major Issues Pertaining to Comprehensively Promoting the Rule of Law,* 140–147. Beijing: People's Publishing House.

Lu, Fang, and Ming Cheng. 2016. "Public Government and the Implicit Size of Government: Based on the Differences between China and the United States." *Journal of Shanghai Administration Institute* 17 (6): 64–77.

Ma, Qingyu, and Xijin Jia. 2015. "The Development Direction and Future Trend of Chinese Social Organization." *Journal of China National School of Administration*, no. 4: 62–67.

Ma, Damien. 2014. "The One-Year Plan." *Foreign Policy* (January): 3.

Ma, Li, Huimin Li, and Qi Ye. 2012. "Analysis of Policy Making Process of China's Energy-Saving Performance Assessment Institution: Taking a Perspective of Central-Local Interaction." *Journal of Public Management* 9 (1): 1–8, 121.

Magill, Bobby. 2016. "Obama Halts Federal Coal Leasing Citing Climate Change." *Scientific American*, January 15.

Mattoo, Aaditya, and Arvind Subramanian. 2012. "Equity in Climate Change: An Analytical Review." *World Development* 40 (6): 1083–1097.

Mertha, Andrew. 2009. "'Fragmented Authoritarianism 2.0': Political Pluralization in the Chinese Policy Process." *China Quarterly* 200 (December): 995–1012.

Mervis, Jeffrey. 2012. "Why NOAA Is in the Commerce Department." *Science Magazine*, January 13. http://www.sciencemag.org/news/2012/01/why-noaa-commerce-department. Accessed July 3, 2018.

Miller, Gary J. 2005. "The Political Evolution of Principal-Agent Models." *Annual Review of Political Science* 8:203–225.

Ministry of Foreign Affairs of China. 2012. "Work Together for a Bright Future of China-US Cooperative Partnership." February 16. http://www.fmprc.gov.cn/mfa_eng/wjdt_665385/zyjh_665391/t910351.shtml. Accessed July 3, 2018.

Molina, Maggie, Patrick Kiker, and Seth Nowak. 2016. *The Greatest Energy Story You Haven't Heard*. Washington, DC: American Council for an Energy Efficient Economy.

Morris, Edmund. 2001. *Theodore Rex*. New York: Modern Library.

Mosendz, Polly. 2016. "What This Election Taught Us about Millennial Voters." *Bloomberg News*, November 9.

Myslikova, Z., K. S. Gallagher, and F. Zhang. 2017. "Mission Innovation 2.0: Recommendations for the Second Mission Innovation Ministerial in Beijing, China." CIEP Climate Policy Lab Discussion Paper 14, The Fletcher School, Tufts University, Medford, MA, May. https://sites.tufts.edu/cierp/files/2017/09/CPL_MissionInnovation014_052317v2low.pdf.

Nagourney, Adam. 2017. "California Extends Climate Bill, Handing Gov. Jerry Brown a Victory." *New York Times*, July 17.

Narassimhan, Easwaran, Kelly S. Gallagher, Stefan Koester, and Julio Rivera Alejo. 2017. "Carbon Pricing in Practice: A Review of the Evidence." CIEP, The Fletcher School, Tufts University, Medford, MA. https://sites.tufts.edu/cierp/files/2017/11/Carbon-Pricing-In-Practice-A-Review-of-the-Evidence.pdf.

National Bureau of Statistics of China. 2015. *China Statistical Yearbook 2015: Divisions of Administrative Areas in China*. Beijing: China Statistics Press.

National Bureau of Statistics of China. 2016. *China Statistical Yearbook 2016*. Beijing: China Statistics Press.

National Development and Reform Commission (NDRC). 2006. "Notice Regarding Top-1,000 Enterprises Energy Conservation Program, No. 571." Beijing: NDRC. http://www.ndrc.gov.cn/rdzt/jsjyxsh/200604/t20060413_66114.html.

National Energy Administration (NEA). 2015. "The Whole Society Electricity Consumption Data for 2014." National Energy Administration, January 16. http://www.nea.gov.cn/2015-01/16/c_133923477.htm.

National Energy Administration (NEA). 2017. "Wind Power Grid Operation in 2016." National Energy Administration, January 26. http://www.nea.gov.cn/2017-01/26/c_136014615.htm.

NDRC (National Development and Reform Commission, China). 2007. *China's National Climate Change Programme 2007*. Beijing: National Development and Reform Commission.

NDRC (National Development and Reform Commission, China). 2009. *China's Policies and Actions for Addressing Climate Change: The Progress Report*. Beijing: National Development and Reform Commission.

NDRC (National Development and Reform Commission, China). 2012. *Second National Communication on Climate Change of the People's Republic of China*. Beijing: National Development and Reform Commission.

NDRC (National Development and Reform Commission, China). 2013. *China's Policies and Actions on Climate Change 2013*. Beijing: National Development and Reform Commission.

NDRC (National Development and Reform Commission, China). 2014. *National Climate Change Plan for 2014–2020*. Beijing: National Development and Reform Commission.

NDRC (National Development and Reform Commission, China). 2016. "The 13th Five Year Plan for the Development of Renewable Energy." National Development and Reform Commission, December 10. https://chinaenergyportal.org/en/13th-fyp-development-plan-renewable-energy/.

NDRC (National Development and Reform Commission, China). 2017a. "Main Functions of the NDRC." National Development and Reform Commission. http://en.ndrc.gov.cn/mfndrc/. Accessed August 23, 2017.

NDRC (National Development and Reform Commission, China). 2017b. "National Leading Group to Address Climate Change." http://qhs.ndrc.gov.cn/ldxz/. Accessed July 3, 2018.

NDRC (National Development and Reform Commission, China). 2017c. "Notice on Printing and Distributing the Implementation Plan of Energy—Saving Action of 1000 Enterprises [Circular 571]." National Development and Reform Commission. http://bgt.ndrc.gov.cn/zcfb/200604/t20060414_499304.html. Accessed September 12, 2017.

NDRC, MOE, MIIT, MOF, MOHURD, MOT, MOFCOM, SASAC, AQSIQ, NBC, CBRC, and NEA. 2011. "Notice of Issuing the 10,000 Enterprises Energy Conservation and Low Carbon Program Implementation Plan." http://bgt.ndrc.gov.cn/zcfb/201112/t20111229_498695.html.

New York City Government. 2014. "One City, Built to Last: Transforming New York City's Buildings for a Low-Carbon Future." http://www.nyc.gov/builttolast.

New York City Government. 2016. "NYC CoolRoofs. https://www1.nyc.gov/nycbusiness/article/nyc-coolroofs.

NPC (National People's Congress). 2000. "Legislation Law of the People's Republic of China." http://en.sunlawyers.com/CN/Law_Order/Show_13_1.html.

NPC (National People's Congress). 2004. "Constitution of the People's Republic of China." As released in 1982, and amended in 1988, 1993, 1999, 2004 respectively. http://www.npc.gov.cn/englishnpc/Constitution/node_2825.htm.

NREL (National Renewable Energy Laboratory). 2016. "Renewable Resources Maps and Data." US Department of Energy. https://www.nrel.gov/gis/mapsearch/.

Obama, Barack. 2008. "Videotaped Remarks to the Bi-Partisan Governors Global Climate Summit." Los Angeles, CA, November 18. Gerhard Peters and John T. Woolley, *The American Presidency Project*. http://www.presidency.ucsb.edu/ws/?pid=84875.

Office of the Chief Financial Officer. 2017. *FY 2018 Budget Justification*. DOE/CF-0134.Washington, DC: DOE.

Office of the Historian, and Office of Art & Archives, Office of the Clerk. n.d. "Power of the Purse." http://history.house.gov/Institution/Origins-Development/Power-of-the-Purse/. Accessed March 5, 2018.

Office of the Press Secretary. 2015. "Factsheet: The White House Releases New Strategy for American Innovation, Announces Areas of Opportunity from Self-Driving Cars to Smart Cities." White House of President Barack Obama, October 21.

Office of the Press Secretary of the White House. 2015. "Fact Sheet: Launching a Public-Private Partnership to Empower Climate-Resilient Developing Nations." White House Office of the Press Secretary, June 9. https://obamawhitehouse.archives.gov/the-press-office/2015/06/09/fact-sheet-launching-public-private-partnership-empower-climate-resilien.

Oliver, Hongyan H., Kelly Sims Gallagher, Donglian Tian, and Jinhua Zhang. 2009. "China's Fuel Economy Standards for Passenger Vehicles: Rationale, Policy Process, and Impacts." *Energy Policy* 37 (11): 4720–4729.

Olivier, Jos G. J., K. M. Schwe, and Jeroen A. H. W. Peters. 2017. *Trends in Global CO2 Emissions: 2017 Report.* The Hague: PBL Netherlands Environmental Assessment Agency; Ispra: European Commission, Joint Research Center.

Oreskes, Naomi. 2004. "The Scientific Consensus on Climate Change." *Science* 306 (5702): 1686.

Oreskes, Naomi, and Erik Conway. 2010. *Merchants of Doubt.* New York: Bloomsbury Publishing.

PBS Frontline. 2007. "An Interview with Senator Chuck Hagel." April 5. https://www.pbs.org/wgbh/pages/frontline/hotpolitics/interviews/hagel.html.

People's Daily. 2016. "Wang Yilin Ren Zhong Petroleum Chairman, Party Secretary." *People's Daily,* May 5. http://www.gov.cn/xinwen/2015-05/05/content_2856866.htm.

People's Daily. 2017. "Taking the Power Generation Industry as a Breakthrough, the National Carbon Emissions Trading System Was Officially Launched." *People's Daily,* December 20. http://www.gov.cn/xinwen/2017-12/20/content_5248688.htm.

Podesta, John, C. H. Tung, Samuel R. Berger, and Jisi Wang. 2013. *Toward a New Model of Major Power Relations.* Washington, DC: Center for American Progress.

Political Bureau of the 18th CPC Central Committee. 2013. "The Decision of the Central Committee of the CPC ('CCCPC') on Some Major Issues Concerning Comprehensively Deepening the Reform." Beijing.

Popovich, Nadja, John Schwartz, and Tatiana Scholossburg. 2017. "How Americans Think about Climate Change, in Six Maps." *New York Times,* March 21.

Price, Lynn, Mark D. Levine, Nan Zhou, David Fridley, Nathaniel Aden, Hongyou Lu, Michael McNeil, Nina Zheng, Yining Qin, and Ping Yowargana. 2011. "Assessment of China's Energy-Saving and Emission-Reduction Accomplishments and Opportunities during the 11th Five Year Plan." *Energy Policy* 39 (4): 2165–2178.

Qi, Liyan, and Te-Ping Chen. 2017. "Chinese Scientist Blasts Trump's Climate-Change Talk." *Wall Street Journal,* January 27. https://blogs.wsj.com/chinarealtime/2017/01/27/0127cpollute/.

Qi, Ye, Li Ma, Huanbo Zhang, and Huimin Li. 2008. "Translating a Global Issue into Local Priority: China's Local Government Response to Climate Change." *Journal of Environment & Development* 17 (4): 379–400.

Qi, Ye, and Huimin Li. 2011. "China's Low Carbon Development under the Eleventh Five Year Plan." [In Chinese.] *China Population, Resources, and Environment* 21 (10): 60–68.

Qi, Ye, and Xiliang Zhang. 2016. *China's Low Carbon Development Report (2015–2016)*. [In Chinese.] Beijing: Social Science Academic Press.

Rabe, Barry George. 2004. *Statehouse and the Greenhouse: The Emerging Politics of American Climate Change Policy*. Washington, DC: Brookings Institution Press.

RGGI (Regional Greenhouse Gas Initiative). 2016. "Fact Sheet: The Investment of RGGI Proceeds through 2014." https://www.rggi.org/sites/default/files/Uploads/Proceeds/RGGI_Proceeds_Report_2014.pdf.

Ru, Peng, Qiang Zhi, Fang Zhang, Xiaotian Zhong, Jianqiang Li, and Jun Su. 2012. "Behind the Development of Technology: The Transition of Innovation Modes in China's Wind Turbine Manufacturing Industry." *Energy Policy* 43:58–69.

Russell, Cristine. 2013. "To Tell a Complicated Climate Science Story: Simplify, Shorten, List." *Columbia Journalism Review*, September 30. http://archives.cjr.org/the_observatory/ipcc_coverage.php.

Saich, Tony. 2011. *Governance and Politics of China*. 3rd ed. New York: Palgrave Macmillan.

Samuelson, Darren. 2009. "Climate Bill Needed to 'Save Our Planet,' Says Obama." *New York Times*, February 25.

Schein, Edgar. 1996. *Strategic Pragmatism: The Culture of Singapore's Economic Development*. Cambridge, MA: MIT Press.

Schmiegelow, Michèle, and Henrik Schmiegelow. 1989. *Strategic Pragmatism: Japanese Lessons in the Use of Economic Theory*. Santa Barbara, CA: Praegar.

SEIA (Solar Energy Industries Association). 2017. "Fact Sheet: Solar Investment Tax Credit (ITC)." Solar Energy Industries Association. https://www.seia.org/research-resources/solar-investment-tax-credit-itc-101.

Selin, H., and S. D. VanDeveer. 2009. "Changing Climates and Institution Building across the Continent." In *Changing Climates in North American Politics: Institutions, Policymaking, and Multilevel Governance*, ed. H. Selin and S. D. VanDeever, 3–22. Cambridge, MA: MIT Press.

Shan, Yuli, Dabo Guan, Jianghua Liu, Zhu Liu, Jingru Liu, Heike Schroeder, Yang Chen, Shuai Shao, Zhifu Mi, and Qiang Zhang. 2016. "CO_2 Emissions Inventory of Chinese Cities." *Atmospheric Chemistry and Physics Discussion*, March 1.

Shenzhen People's Government. 2014. "Decree of the Shenzhen Municipal People's Government (no. 262): Interim Measures of Shenzhen for Administration of Carbon Emission Trading." Shenzhen People's Government. http://www.sz.gov.cn/zfgb/2014/gb876/201404/t20140402_2335498.htm.

Shi, Hexing. 2014. "The People's Congress System and China's Constitutional Development." In *China's Political Development: Chinese and American Perspectives*, ed. Kenneth G. Lieberthal, Cheng Li, and Keping Yu, 103–120. Washington, DC: Brookings Institution Press.

Shirk, Susan. 2011. *Changing Media, Changing China.* Oxford: Oxford University Press.

Smith, Kirk, and Evan Haigler. 2008. "Co-benefits of Climate Mitigation and Health Protection in Energy Systems: Scoping Methods." *Annual Review of Public Health* 29:11–25.

State Council. 2003. "Order Number 369 of the Regulations on the Administration of Levy and Use of Sewage Charge." http://www.gov.cn/zhengce/content/2008-03/28/content_5152.htm.

State Council. 2005a. "Notice of the State Council on Suggestions on the Developing of the Circular Economy." http://www.gov.cn/zwgk/2005-09/08/content_30305.htm.

State Council. 2005b. "Several Opinions of the State Council on Speeding Up the Development of Circular Economy." http://www.gov.cn/zwgk/2005-09/08/content_30305.htm.

State Council. 2010. "Notice of the State Council on Establishing the Food Safety Committee of the State Council." http://www.gov.cn/zwgk/2010-02/10/content_1532419.htm.

State Council. 2011a. *12th Five-Year Work Plan for Controlling Greenhouse Gas Emissions.* Beijing: State Council.

State Council. 2011b. *Comprehensive Work Plan on Energy Conservation and Emissions Reduction for the 12th Five Year Plan Period.* Beijing: State Council.

State Council. 2011c. "Part VI: Building Resource Conservative and Environmentally Friendly Society." In *The 11th Five Year Plan for National Economic and Social Development.* Bejing: State Council.

State Council. 2011d. "The Work Plan for Greenhouse Gas Emission Control during the 12th Five-Year Plan Period." The State Council [online database]. Beijing.

State Council. 2012a. *Energy Saving and New Energy Vehicle Industry Development Plan.* Beijing: State Council.

State Council. 2012b. "Notice of the State Council on Issuing the Planning for the Development of the Energy-Saving and New Energy Automobile Industry (2012–2020)." The State Council [online database]. Beijing.

State Council. 2015. *Notice of the State Council on Printing and Distributing the Made in China 2025*. Beijing: State Council.

State Council. 2016a. *13th Five-Year Work Plan for Controlling Greenhouse Gas Emissions*. Beijing: State Council.

State Council. 2016b. *Comprehensive Work Plan on Energy Conservation and Emissions Reduction for the 13th Five Year Plan Period*. Beijing: State Council.

State Council. 2016c. *Notice of the State Council on Printing and Distributing the Work Plan for Controlling Greenhouse Gas Emission in the 13th Five-Year Plan*. Beijing: State Council.

Stern, Todd. 2015. "Testimony of Special Envoy for Climate Change Todd D. Stern at the U.N. Climate Change Conference in Paris." Paris. https://www.foreign.senate.gov/imo/media/doc/102015_Stern_Testimony.pdf.

Stocker, T. F., D. Qin, G.-K. Plattner, L. V. Alexander, S. K. Allen, N. L. Bindoff, F.-M. Bréon, et al. 2013. "Technical Summary." In *Climate Change 2013: The Physical Science Basis: Contribution of Working Group I to the Fifth Assessment Report of the Intergovernmental Panel on Climate Change*, ed. T. F. Stocker, D. Qin, G.-K. Plattner, M. Tignor, S. K. Allen, J. Boschung, A. Nauels, et al., 55. Cambridge: Cambridge University Press.

Stocking, Andrew, and Terry Dinan. 2015. "China's Growing Energy Demand: Implications for the United States." Working Paper 20150-5. Washington, DC: Congressional Budget Office.

Stohr, Greg, and Jennifer Dlouhy. 2016. "Obama's Clean Power Plan Put on Hold by U.S. Supreme Court." *Bloomberg*, February 9.

Stone, Deborah. 1989. "Causal Stories and the Formation of Policy Agendas." *Political Science Quarterly* 104 (2): 281–300.

Supreme Court of the United States. 2010. "Citizens United v. Federal Election Commission." 558 US 310, Washington, DC. https://transition.fec.gov/law/litigation/cu_sc08_opinion.pdf.

Supreme People's Court of China (SPC). 2017a. "Leaders of the Supreme People's Court of the People's Republic of China." http://www.court.gov.cn/jigou-fayuan lingdao.html.

Supreme People's Court of China (SPC). 2017b. "The Functions and Rights of the Supreme People's Court." http://english.court.gov.cn/2015-07/16/content _21299713.htm.

Teng, Fei. 2015. "Paris Agreement: Success or Failure." [In Chinese.] *China Reform,* December 1.

Tocqueville, Alexis de. [1840] 2006. *Democracy in America, Volume 1.* Translated by Henry Reeve. http://www.gutenberg.org/files/815/815-h/815-h.htm#link2HCH0051.

Trembath, Alex, Jesse Jenkins, Ted Nordhaus, and Michael Shellenberger. 2012. *Where the Shale Gas Revolution Came From: Government's Role in the Development of Hydraulic Fracturing in Shale.* Oakland, CA: Breakthrough Institute.

Trump, Donald J. 2012. "The concept of global warming was created by and for the Chinese in order to make U.S. manufacturing non-competitive." Twitter, November 6, 2012, 2:15 PM. https://twitter.com/realDonaldTrump/status/265895292191248385.

Trump, Donald J. 2017. "The Inaugural Address." White House, January 20. https://www.whitehouse.gov/inaugural-address.

UN Climate Change Newsroom. 2014. "U.S. China Climate Moves Boost Paris Prospects." *UN Climate Change Newsroom,* November 12.

UNFCCC (United Nations Framework Convention on Climate Change). 1998. "Kyoto Protocol to the UN Framework Convention on Climate Change." United Nations. http://unfccc.int/resource/docs/convkp/kpeng.pdf.

UNFCCC (United Nations Framework Convention on Climate Change). 2016a. "NDC Registry." http://www4.unfccc.int/ndcregistry/Pages/Home.aspx.

UNFCCC (United Nations Framework Convention on Climate Change). 2016b. "UN Updates Synthesis Report of National Climate Plans." *UN Climate Change Newsroom,* May 2. http://newsroom.unfccc.int/unfccc-newsroom/synthesis-report-update-of-national-climate-plans/.

UNFCCC (United Nations Framework Convention on Climate Change). 2017a. "Essential Background of the UNFCCC." http://unfccc.int/files/essential_background/background_publications_htmlpdf/application/pdf/conveng.pdf.

UNFCCC (United Nations Framework Convention on Climate Change). 2017b. "Status of Ratification of the Convention." http://unfccc.int/essential_background/convention/status_of_ratification/items/2631.php.

United Nations. 1997. "Institutional Aspects of Sustainable Development in China." United Nations Commission on Sustainable Development, April 1. http://www.un.org/esa/agenda21/natlinfo/countr/china/inst.htm.

US Census Bureau. 2016. *New Census Data Show Differences between Urban and Rural Populations,* B16–B210. Washington, DC: US Census Bureau.

US Code. 2017a. "United States Code Title 42, Chapter 85, Subchapter I, Part A, Sec. 7409: National Primary and Secondary Ambient Air Quality Standards." https://

www.gpo.gov/fdsys/pkg/USCODE-2013-title42/html/USCODE-2013-title42-chap85 -subchapI-partA-sec7409.htm. Accessed December 31, 2017.

US Code. 2017b. "United States Code Title 42, Chapter 85, Subchapter II, Part A, Sec. 7521: Emission Standards for New Motor Vehicles or New Motor Vehicle Engines." https://www.law.cornell.edu/uscode/text/42/7521. Accessed December 31, 2017.

US Court of Appeals. 2017. "Clean Air Council v. Scott Pruitt." No. 17-1145 (District of Columbia Circuit 2017). https://www.cadc.uscourts.gov/internet/opinions.nsf/ a86b20d79beb893e85258152005ca1b2/$file/17-1145-1682465.pdf.

US Department of State. 2016. *2016 Second Biennial Report of the United States of America under the United Nations Framework Convention on Climate Change*. Washington, DC: US Department of State.

US Justice Department. 2018. "Citizens Guide to U.S. Federal Law on Child Pornography." https://www.justice.gov/criminal-ceos/citizens-guide-us-federal-law-child -pornography. Accessed July 3, 2018.

US Office of Personnel Management. n.d. "Historical Federal Workforce Tables: Total Government Employment since 1962." https://www.opm.gov/policy-data -oversight/data-analysis-documentation/federal-employment-reports/historical -tables/executive-branch-civilian-employment-since-1940/. Accessed March 5, 2017.

US Senate. 2017. "Senate Committees." US Senate website. https://www.senate.gov/ artandhistory/history/common/briefing/Committees.htm.

Van Natta, Don, and Neela Banerjee. 2002. "Top GOP Donors in Energy Industry Met Cheney Panel." *New York Times*, March 1.

Vidal, John. 2005. "Revealed: How Oil Giant Influenced Bush." *Guardian*, June 8.

Wall Street Journal. 2008. "Rahm Emanuel on the Opportunities of Crisis." *Wall Street Journal*, YouTube Channel, November 19. https://www.youtube.com/watch ?v=_mzcbXi1Tkk.

Wang, Alex, and Gao Jie. 2010. "Environmental Courts and the Development of Environmental Public Interest Litigation in China." *Journal of Court Innovation* 3:37.

Wang, Lu. 2015. "New Policy to Target Industrial Sector Coal Consumption." *Economic Information Daily (Beijing)*, March 2.

Wang, Yiming. 2017. "Remarks at Clean Energy Ministerial." Development Research Center of the State Council, Beijing, June 6.

Wang, Youjuan, Weibin Lin, and Qian Wan. 2013. "Green Growth: Constructing a Resource-Saving and Environment-Friendly Production Pattern." In *China Green Development Index Report 2011*, ed. X. Li and J. Pan, 31–48. Berlin: Springer-Verlag.

Weible, Christopher. 2014. "Introduction: The Scope and Focus of Policy Process Research and Theory." In *Theories of the Policy Process*, 3rd ed., ed. Paul A. Sabatier and Christopher M. Weible, 3–22. Boulder, CO: Westview Press.

White House. 2012a. "Obama Administration Finalizes Historic 54.5 MPG Fuel Efficiency Standards." August 28. https://obamawhitehouse.archives.gov/the-press -office/2012/08/28/obama-administration-finalizes-historic-545-mpg-fuel -efficiency-standard.

White House. 2012b. "Remarks by Vice President Biden and Chinese Vice President Xi at the State Department Luncheon." February 14. https://obamawhitehouse .archives.gov/the-press-office/2012/02/14/remarks-vice-president-biden-and-chinese -vice-president-xi-state-departm.

White House. 2013. "President Obama's Climate Action Plan." June 25. https:// obamawhitehouse.archives.gov/the-press-office/2013/06/25/fact-sheet-president -obama-s-climate-action-plan.

White House. 2014. "U.S.-China Joint Announcement on Climate Change." November 11. https://obamawhitehouse.archives.gov/the-press-office/2014/11/11/us-china -joint-announcement-climate-change.

Wike, Richard. 2016. "6 Facts about How Americans and Chinese See Each Other." Pew Research Center [database online]. Washington, DC.

Wike, Richard, Bruce Stokes, and Jacob Poushter. 2015. *Global Publics Back U.S. on Fighting ISIS, but Are Critical of Post-9/11 Torture*. Washington, DC: Pew Research Center.

Wildau, Gabriel. 2016. "China Moves to De-politicise Management of State-Owned Enterprises." *Financial Times*, July 3.

Wiser, Ryan, and Mark Bollinger. 2017. "2016 Wind Technologies Market Report." https://emp.lbl.gov/sites/default/files/2016_wtmr_data_file.xls.

Wiser, Ryan H., Mark Bolinger, and Galen L. Barbose. 2007. "Using the Federal Production Tax Credit to Build a Durable Market for Wind Power in the United States." *Electricity Journal* 20 (9): 77–88.

World Bank. 2009. *World Development Report: Public Attitudes toward Climate Change: Findings from a Multi-country Poll*. Washington, DC: World Bank.

World Bank. 2017. *World Development Indicators*. Washington, DC: World Bank.

WRI (World Resources Institute). 2014. "6 Graphs Explain the World's Top-10 Emitters." WRI Blogspot, November 25. https://wwwri.org/blog/2014/11/6-graphs -explain-world's-top-10-emitters.

WRI (World Resources Institute). 2017. "CAIT (the Climate Access Indicators Tool) Climate Data Explorer—Historical Emissions." http://cait.wri.org/historical.

Wu, Yuwen. 2015. "Profile: China's Fallen Security Chief Zhou Yongkang." *BBC News*, October 12.

Xi, Jinping. 2012. "Work Together for a Bright Future of China-US Cooperative Partnership." Speech, February 16. www.china-embassy.org/eng/zgyw/t910351.htm.

Xi, Jinping. 2014. *The Governance of China*. 1st ed. Beijing: Foreign Languages Press.

Xi, Jinping. 2017. "Report at the 19th Party Congress." Translated by Xinhua. http://www.xinhuanet.com/english/download/Xi_Jinping's_report_at_19th_CPC_National_Congress.pdf.

Xinhua. 2006. "China Issues S&T Development Guidelines." *Xinhua News Agency*, February 9.

Xinhua. 2013. "Decision of the Central Committee of Communist Party of China on Some Major Issues Concerning Comprehensively Deepening the Reform." *Xinhua News Agency*, November 15.

Xinhua. 2015a. "'Made in China 2025' Plan Unveiled." *Xinhua News Agency*, May 19.

Xinhua. 2015b. "Recommendations for the 13th Five-Year Plan for Economic and Social Development." *Xinhua News Agency*, November 3.

Xinhua. 2015c. "South-South Cooperation Vital to Global Action against Climate Change." *Xinhua News Agency*, December 7.

Xinhua. 2015d. "The Proposal of the CPC Central Committee on the Formulation of the 13th Five Year Plan." *Xinhua News Agency*, November 3. http://www.gov.cn/xinwen/2015-11/03/content_2959432.htm.

Xinhua. 2015e. "China Unveils New, Ambitious Climate Goals." *Xinhua News Agency*, July 1.

Xinhua. 2016a. "The 13th Five-Year Plan for the National Economic and Social Development of the People's Republic of China." *Xinhua News Agency*, March 17. http://www.gov.cn/xinwen/2016-03/17/content_5054992.htm.

Xinhua. 2016b. "The Standing Committee of the National People's Congress Releases Legislative Work Plan for 2016." *Xinhua News Agency*, April 22.

Xinhua. 2017a. "China Focus: BASIC Countries Make 'Important Contribution' to Paris Agreement." *Xinhua News Agency*, April 11.

Xinhua. 2017b. "Full Text of Xi Jinping's Report at 19th CPC National Congress." *Xinhua News Agency*, November 3. http://news.xinhuanet.com/english/special/2017-11/03/c_136725942.htm.

Xinhua. 2017c. "Pilots the Strong Leader in the New Era: A Record of the New Central Leadership of the Party." *Xinhua News Agency*, October 26. http://news.xinhuanet.com/politics/19cpcnc/2017-10/26/c_1121860147.htm.

Yang, Dali. 2017. "China's Illiberal Regulatory State." *China Political Science Review* 2:114–133.

Yang, Guangbin. 2014. "Decentralization and Central-Local Relations in Reform-Era China." In *China's Political Development: Chinese and American Perspectives*, ed. Kenneth G. Lieberthal, Cheng Li, and Keping Yu, 254–281. Washington, DC: Brookings Institution Press.

Yang, Wanli. 2015. "Rural-Urban Income Gap Narrows." *China Daily*, April 22.

Yilin, Little. 2016. "Renewable Energy Law Ten Years of Reflection: The Old Model Is Difficult to Follow with the Rapid Expansion of Subsidies Gap." *Solar OFWeek*, September 14. http://solar.ofweek.com/2016-09/ART-260009-8420-30038530.html.

Yin, Zhongqiang. 2015. "The Main Problems in the Carbon Trade Pilot in China and the Challenge for the National Carbon Market Establishment." [In Chinese.] *Low Carbon World*, October 12.

Yu, Keping, ed. 2010. *Democracy and the Rule of Law in China*. Leiden: Brill.

Yu, Keping. 2011. "Civil Society in China: Concepts, Classification, and Institutional Environment." In *State and Civil Society: The Chinese Perspective*, ed. Deng Zhenlai, 63–96. Singapore: World Scientific.

Yu, Keping. 2014. "The People's Republic of China's Sixty Years of Political Development." In *China's Political Development: Chinese and American Perspectives*, ed. Kenneth G. Lieberthal, Cheng Li, and Keping Yu, 39–61. Washington, DC: Brookings Institution Press.

Yu, Keping. 2016. *Democracy in China: Challenge or Opportunity*. Singapore: World Scientific Publishing Co.

Yu, Xiang, and Alex Y. Lo. 2015. "Carbon Finance and the Carbon Market in China." *Nature Climate Change* 5:15–16.

Yuan, D., and J. Feng. 2011. "Behind China's Green Goals." *China Dialogue*, March 24.

Zahariadis, Nikolaos. 2014. "Ambiguity and Multiple Streams." In *Theories of the Policy Process*, 3rd. ed., ed. Paul A. Sabatier and Christopher M. Weible, 25–57. Boulder, CO: Westview Press.

Zartman, I. William. 2000. "Ripeness: The Hurting Stalemate and Beyond." In *International Conflict Resolution after the Cold War*, ed. Paul Stern and Daniel Druckman, 225–250. Washington, DC: National Academies Press.

Zeng, Jinghan. 2016. "Constructing a 'New Type of Great Power Relations': The State of Debate in China (1998–2014)." *British Journal of Politics and International Relations* 18 (2): 422–442.

Zhang, Da, Valerie J. Karplus, Cyril Cassisa, and Xiliang Zhang. 2014. "Emissions Trading in China: Progress and Prospects." *Energy Policy* 75:9–16.

Zhang, Fang, and Keman Huang. 2017. "The Role of Government in Industrial Energy Conservation in China: Lessons from the Iron and Steel Industry." *Energy for Sustainable Development* 39:101–114.

Zhang, Fang, and Kelly S. Gallagher. 2016. "Innovation and Technology Transfer through Global Value Chains: Evidence from China's PV Industry." *Energy Policy* 94:191–203.

Zhao, Lifeng, and Kelly S. Gallagher. 2007. "Research, Development, Demonstration, and Early Deployment Policies for Advanced-Coal Technology in China." *Energy Policy* 35 (12): 6467–6477.

Zhao, Suisheng, ed. 2016. *Chinese Foreign Policy: Pragmatism and Strategic Behavior.* New York: Routledge.

Zhao, Xiaoni, and Baiyu Zhuang. 2016. "The Third National Expert Committee on Climate Change Was Established and First Meeting Held." *China Meteorological News Press*, September 30.

Zhao, Xingshu, 2017. "Trump Administration and the Compliance of the Paris Agreement." [In Chinese.] *Climate Change Research* 13 (5): 448–455.

Zheng, Yongnian. 2010. *The Chinese Communist Party as Organizational Emperor: Culture, Reproduction, and Transformation.* London: Routledge.

Zhou, Guanghui. 2014. "Contemporary China's Decisionmaking System." In *China's Political Development: Chinese and American Perspectives*, ed. Kenneth G. Lieberthal, Cheng Li, and Keping Yu, 340–358. Washington, DC: Brookings Institution Press.

Zhou, Nan, Mark Levine, and Lynn Price. 2010. "Overview of Current Energy Efficiency Policies in China." *Energy Policy* 38 (11) (November): 6439–6452.

Zhou, Sheng, and Xiliang Zhang. 2010. "Nuclear Energy Development in China: A Study of Opportunities and Challenges." *Energy Policy* 35 (11): 4282–4288.

Zhu, Guanglei. 2008. *The Process of Contemporary Chinese Government.* Tianjin: Tianjin People's Publishing House.

Zhu, Guanglei, and Zhihong Zhang. 2005. "A Critique of Isomorphic Responsibility Governmental System." *Journal of Peking University (Philosophy and Social Science)* 42 (1): 101–112.

Zhu, Junqing. 2016. "Opinion: Trump's Rise Is Fall of U.S. Democracy." *New China*, March 4. http://news.xinhuanet.com/english/2016-03/04/c_135156215.htm.

Zhuang, Guiyang. 2009. "The Barriers to a Low-Carbon Economy in China." *Jiangxi Social Sciences Journal*, no. 7: 20–26.

Zou, Ji. 2014. "Four Reasons Why the China-US Climate Statement Matters." *China Dialogue*, December 12. https://www.chinadialogue.net/blog/7495-Four-reasons-why -the-China-US-climate-statement-matters/en.

Zou, Ji, Xiaohua Zhang, Sha Fu, Yue Qi, Ji Chen, and Hairan Gao. 2014. "Implications and Challenges of the U.S.-China Joint Announcement on Climate Change Cooperation." *China Carbon Forum*. http://www.chinacarbon.info/wp-content/ uploads/2014/11/Implications-of-the-US-China-Joint-Announcement-on-Climate -Change_%E4%B8%AD%E5%9B%BD%E7%A2%B3%E8%AE%BA%E5%9D%9B.pdf.

Index

American and Comparative Environmental Policy

Sheldon Kamieniecki and Michael E. Kraft, series editors

Russell J. Dalton, Paula Garb, Nicholas P. Lovrich, John C. Pierce, and John M. Whiteley, *Critical Masses: Citizens, Nuclear Weapons Production, and Environmental Destruction in the United States and Russia*

Daniel A. Mazmanian and Michael E. Kraft, editors, *Toward Sustainable Communities: Transition and Transformations in Environmental Policy*

Elizabeth R. DeSombre, *Domestic Sources of International Environmental Policy: Industry, Environmentalists, and U.S. Power*

Kate O'Neill, *Waste Trading among Rich Nations: Building a New Theory of Environmental Regulation*

Joachim Blatter and Helen Ingram, editors, *Reflections on Water: New Approaches to Transboundary Conflicts and Cooperation*

Paul F. Steinberg, *Environmental Leadership in Developing Countries: Transnational Relations and Biodiversity Policy in Costa Rica and Bolivia*

Uday Desai, editor, *Environmental Politics and Policy in Industrialized Countries*

Kent Portney, *Taking Sustainable Cities Seriously: Economic Development, the Environment, and Quality of Life in American Cities*

Edward P. Weber, *Bringing Society Back In: Grassroots Ecosystem Management, Accountability, and Sustainable Communities*

Norman J. Vig and Michael G. Faure, editors, *Green Giants? Environmental Policies of the United States and the European Union*

Robert F. Durant, Daniel J. Fiorino, and Rosemary O'Leary, editors, *Environmental Governance Reconsidered: Challenges, Choices, and Opportunities*

Paul A. Sabatier, Will Focht, Mark Lubell, Zev Trachtenberg, Arnold Vedlitz, and Marty Matlock, editors, *Swimming Upstream: Collaborative Approaches to Watershed Management*

Sally K. Fairfax, Lauren Gwin, Mary Ann King, Leigh S. Raymond, and Laura Watt, *Buying Nature: The Limits of Land Acquisition as a Conservation Strategy, 1780-2004*

Steven Cohen, Sheldon Kamieniecki, and Matthew A. Cahn, *Strategic Planning in Environmental Regulation: A Policy Approach That Works*

Michael E. Kraft and Sheldon Kamieniecki, editors, *Business and Environmental Policy: Corporate Interests in the American Political System*

Joseph F. C. DiMento and Pamela Doughman, editors, *Climate Change: What It Means for Us, Our Children, and Our Grandchildren*

Christopher McGrory Klyza and David J. Sousa, *American Environmental Policy, 1990–2006: Beyond Gridlock*

John M. Whiteley, Helen Ingram, and Richard Perry, editors, *Water, Place, and Equity*

Judith A. Layzer, *Natural Experiments: Ecosystem-Based Management and the Environment*

Daniel A. Mazmanian and Michael E. Kraft, editors, *Toward Sustainable Communities: Transition and Transformations in Environmental Policy*, second edition

Henrik Selin and Stacy D. VanDeveer, editors, *Changing Climates in North American Politics: Institutions, Policymaking, and Multilevel Governance*